Graham Brodie, Dorin Gupta, Muhammad Jamal Khan, Sally Foletta, Natalie Bootes

**Microwave Based Weed Control and Soil Treatment**

Graham Brodie, Dorin Gupta, Muhammad Jamal Khan, Sally Foletta, Natalie Bootes

# Microwave Based Weed Control and Soil Treatment

---

Managing Editor: Agnieszka Topolska

DE GRUYTER

ISBN: 978-3-11-060519-8
e-ISBN: 978-3-11-060557-0



**Library of Congress Cataloging-in-Publication Data**
A CIP catalog record for this book has been applied for at the Library of Congress.


Published by De Gruyter Poland, Warsaw/Berlin
Part of Walter de Gruyter GmbH, Berlin/Boston
The book is published with open access at www.degruyter.com.

Managing Editor: Agnieszka Topolska

www.degruyter.com
Cover illustration: © Graham Brodie

# Contents

Preface—1

1 General Introduction—3
1.1 Introduction—3
1.2 Basic Weed Science—3
1.3 The Microwave Option—4
1.4 Reference —4

2 The Growing Threat to Herbicide Use—6
2.1 Introduction—6
2.2 A Brief History of Herbicides—7
2.3 The Development of Herbicide Resistance—8
2.4 Indirect Costs of Herbicide Usage—10
2.5 Conclusion—11
2.6 References —11

3 A System Model for Crop Yield Potential as a Function of Herbicide Weed Control over Time—14
3.1 Introduction—14
3.2 Derivation of Crop System Transfer Function for Herbicide Weed Management —15
3.3 Herbicide Weed Management—15
3.4 Sensitivity Analysis—17
3.5 Examples—19
3.6 Conclusion—19
3.7 References—22

4 Physical Weed Control—24
4.1 Introduction —24
4.2 Tillage—24
4.2.1 Hand Hoeing—25
4.2.2 Mechanical Tillage—25
4.3 Thermal Weed Control—26
4.3.1 Flaming —27
4.3.2 Steam Treatment—28
4.3.3 Hot Water—31
4.4 Radiation Systems—31
4.4.1 Infrared Radiation—32
4.4.2 Ultraviolet Radiation—35
4.4.3 Lasers—36

4.5 Conclusions —— **38**
4.6 References —— **38**
4.7 Appendix A —— **41**
4.8 Appendix B —— **41**

**5 A Brief Review of Microwave Heating —— 43**
5.1 Introduction —— **43**
5.2 Microwave Frequency and its Influence over Microwave Heating —— **44**
5.3 Electric Field Interactions with the Material and Their Influence over Microwave Heating —— **45**
5.4 Challenges Associated with Microwave Research —— **48**
5.5 Non-Thermal Effects —— **51**
5.6 References —— **53**

**6 A Brief History of Microwave Weed Control Research —— 55**
6.1 Introduction —— **55**
6.2 Pioneering Work —— **55**
6.3 Microwave Weed and Soil Treatment Patents —— **60**
6.4 Conclusion —— **61**
6.5 References —— **63**

**7 Applying Microwave Energy to Plants and the Soil —— 66**
7.1 Introduction —— **66**
7.2 Microwave Applicators —— **66**
7.2.1 Wave-Guide Applicators —— **67**
7.2.2 Resonant Applicators —— **69**
7.3 Antennas —— **70**
7.4 Horn Antenna —— **71**
7.5 Soil Temperature —— **74**
7.5.1 Using the Model to Understand Heat Distributions —— **76**
7.6 Novel Design Considerations —— **79**
7.7 Conclusion —— **80**
7.8 References —— **80**
7.9 Appendix A - Derivation of Near Field from a Uniformly Illuminated Rectangular Aperture —— **82**
7.10 Appendix B - Derivation of Near Field from a Tapered Rectangular Aperture —— **83**

**8 The Potential of Microwave Treatment to Kill Weed Plants —— 86**
8.1 Introduction —— **86**
8.2 Early Experimental Prototypes —— **87**
8.2.1 Static Experiments on Plants —— **89**

8.2.2 Moving Trailer Experiments—90
8.3 Conclusion —**92**
8.4 References —**94**

**9 The Potential of Microwave Soil Treatment to Kill Weed Seeds—96**
9.1 Introduction—**96**
9.2 Seed Treatment—**97**
9.3 Modelling Seed Response—**97**
9.4 Treatment Efficiency —**98**
9.5 Conclusion —**100**
9.6 References —**101**
9.7 Appendix - Analysis of Energy in Soil after Microwave Treatment—**102**

**10 The Effect of Microwave Treatment on Soil Biota—106**
10.1 Introduction—**106**
10.2 Effect of Microwave Treatment of Soil Bacteria—**107**
10.3 Assessment of Fungi and other Soil Biota—**111**
10.4 Some Thoughts about Soil Biota—**116**
10.5 Conclusions—**117**
10.6 References—**118**

**11 The Effect of Microwave Soil Treatment on Subsequent Crop Growth and Yield—121**
11.1 Introduction—**121**
11.2 Initial Pot Trials—**121**
11.2.1 Soil Analysis—**124**
11.3 Pot Trials on Wheat with Combined Microwave and Nitrogen Treatments —**125**
11.3.1 An Initial Microwave-Nitrogen Interaction Experiment—**125**
11.3.2 A Repeat Microwave-Nitrogen Interaction Experiment, using the Same Soil—**129**
11.3.3 A Repeat Microwave-Nitrogen Interaction Experiment, using Fresh Soil—**130**
11.4 Field Experiment – Rice at Dookie Campus—**134**
11.5 Field Experiment – Rice at two Agroecological Zones of Australia —**137**
11.6 Field Experiment – Tomato—**139**
11.7 Field Experiment – Wheat (Dookie)—**141**
11.8 General Discussion & Conclusion—**143**
11.9 References —**145**
11.10 Appendices—**147**

**12 A System Model for Crop Yield Potential as a Function of Microwave Weed Control over Time —— 158**
12.1 Introduction —— 158
12.2 Conclusion —— 162
12.3 References —— 163

**13 A Preliminary Economic Assessment of the Microwave Technology in an Herbicide Resistant World —— 164**
13.1 Introduction —— 164
13.2 Microwave Weed Control —— 164
13.3 References —— 167

**14 Industry Acceptance and Conclusion —— 168**
14.1 Introduction —— 168
14.2 Safety —— 168
14.3 Industry Acceptance —— 169
14.4 Key Features of the Technology —— 169
14.5 Summary —— 170
14.6 General Conclusion —— 171
14.7 References —— 172

**Table of Figures —— 174**

**Index —— 179**

# Preface

Herbicide resistance has become an important constraint on modern agricultural practices. Many resistant biotypes withstand exposure to herbicides from different herbicide families targeting the same site of action, a phenomenon known as cross resistance. An alarming increase in weed biotypes resistant to herbicides targeting different sites of action, referred to as multiple resistance, has also been reported. The declining availability of herbicide chemistries over the last 25 years due to: environmental and human health concerns; lack of market; perceived lack of profitability; and the absence of novel herbicide chemical discovery in recent decades has further compounded the situation. Opportunity exists for a novel weed management technology, which is also compatible with no-till agricultural practices. Various alternative weed management technologies are being considered. These include: tillage, flaming, steam treatment, electrocution, electrostatic fields, microwave, infrared, ultraviolet, lasers, and robotics. Many of these show promise; however, microwave treatment of weed plants and soil seems most compatible with no-till agriculture. It can substitute as a knock-down weed plant killer, or be applied to the soil as a pre-sowing soil fumigation treatment.

Microwave frequencies occupy the portion of the electromagnetic spectrum (300MHz to 300GHz) that lies between VHF radio-waves and thermal infrared. Their application falls into two categories, depending on whether the wave is used to transmit information or energy. The first category includes terrestrial and satellite communication links, radar, radio-astronomy, microwave thermography, material permittivity measurements, and so on. The second category of applications is associated with microwave heating and wireless power transmission. In the case of microwave heating, there is usually no signal modulation and the electromagnetic wave interacts directly with solid or liquid materials. Microwave weed management fits into the second category of applications.

When the intensity of the microwave fields are moderate, different species of weed plants, which have already emerged, respond differently to microwave treatment. Generally, broad-leaf weeds are more susceptible to microwave treatment than grasses. These differences in response probably depend on the locations of the apical meristem, which is at the base of the plant in the case of grasses, and the lethal temperature-time curve for the species.

If the microwave field is intense enough, very rapid volumetric heating and some thermal runaway in the structures cause micro-steam explosions in the plant cells rupture the plant structures, which leads to death. These micro-steam explosions circumvent the normal temperature-time response of the plants and may lead to more efficient weed plant control.

Soil treatment requires significantly more energy; however, there are secondary benefits for crops growing in microwave treated soil. These include: significant reduction of the dormant weed seed bank; significant reduction of nematode populations; significant reduction of fungal populations; better availability of indigenous nitrogen for the plants; more rapid humification; and significant increases in crop growth and yield.

Microwave weed plant treatment can be regarded as a “knock-down” technique and, based on preliminary work with a trailer based prototype with four 2 kW microwave generators, the estimated expenditure for this form of weed control may be like that of conventional weed management techniques. Microwave soil treatment is far more energy expensive and should be considered as equivalent to soil fumigation. Both are expensive compared to conventional herbicide weed management. Both provide equivalent crop growth and yield benefits, and preliminary estimates suggest that both cost about the same to implement.

In potted experiments, crop yields from microwave treated soils were consistently between 60% and 170% higher than the untreated controls. Under field conditions, the increase in yield was generally between 30% and 50% higher than the untreated controls. These responses are also dependent on the applied microwave energy dose, and may be linked to changes in the soil biota brought about by the thermal death of some species and the resilience of others.

Microwave weed management and soil treatment is not restricted by weather conditions; therefore, the technology may offer some timeliness and environmental benefits, which are yet to be quantified in a cropping system.

# 1 General Introduction

## 1.1 Introduction

David Pimentel (2005) suggested that 3 billion kg of pesticides were applied to crops, globally. Despite this, it was estimated that 37% of global crop production was lost to pests (Pimentel, 1997). Insects destroy 13%, plant pathogens 12%, and weeds 12% (Pimentel, 1997). Weeds are the major hindrance in crop production. They compete for light, space, nutrients, moisture and $CO_2$ and significantly decline crops yield all over the world. In the United States, the cost of invasive species was estimated to be US$120 billion annually (Chong, Corlett, Yeo, & Tan, 2011). In Australian agricultural industries, the total estimated cost of weed management and loss in crop productivity due to weeds was about AU$4 billion annually (DAFF, 2006). Chemical and mechanical weed control methods are most commonly used in existing cropping systems (Batlla & Benech-Arnold, 2007; Bebawi et al., 2007). Agronomists normally use fire, flaming, grazing, soil fumigants, mechanical eradication, and biocontrol agents in various cropping system to control weeds below the economic threshold level in the absences of weedicides (Gourd, 2002); however, herbicides are becoming the most common method of weed control in production systems. The global herbicide market was estimated to be $23.97 billion in 2016 and is estimated to reach $34.10 billion by 2022 (Research & Markets, 2017).

## 1.2 Basic Weed Science

Seed production is the key element of long-standing weed population dynamics (Davis, Dixon & Liebman, 2003). The average seed production capacity of barnyard grass, the major problematic of rice crop globally, ranged from 20,000 seed plant $^{-1}$ if emerge with the rice to 7,300 seed plant$^{-1}$, if germinates 35 days after rice crop sowing (Bagavathiannan, Norsworthy, Smith, & Neve, 2011). The weed seedbank is constantly dynamic in existing cropping system and the size of the seedbank is regulated by various factors including germination, emergence, fecundity, predation and deterioration (Gallandt, 2006). The size of the weed seedbank of *E. crus-galli* ranges from 0 – 36080 seeds m$^{-2}$ with an average seed density of 1640 m$^{-2}$ (Bagavathiannan et al., 2011).

Weeds are the major biological constraint of agriculture productivity. Oerke (2006) estimated that 34% crop losses globally can be accounted for because of weeds. The dramatic decline in the production of rice due to weeds all over the world is 10% (Oerke & Dehne, 2004 ). The probability of yield loss in direct seeded rice crop is as high as 50 – 91% as compared to transplanted rice, because there is neither a difference in the size of crop and weed plants nor a destructive effect of flooding on weed emergence at the vegetative stage of transplanted rice (Rao, Johnson, Sivaprasad, Ladha, & Mortimer, 2007). Productivity losses of rice crops generally depend on climatic conditions, existing weed species, weed population density, rice variety, growth rate and weed management practices. Barnyard grass (*Echinochloa crus-galli* L.) is the major problematic bio-agent of rice growing areas (Norsworthy, Burgos, Scott, & Smith, 2007) and is also considered to be the main weed of several semi-aquatic cropping systems (Holm, Plucknett, Pancho, & Herberger, 1991). Barnyard grass (*Echinochloa crus-galli* L.) follows the $C_4$ photosynthetic pathway (Rao et al., 2007) and having indistinguishable morphology to rice at seedling stage, is extremely competitive with the rice crop. Reductions in yield of 30 – 100% have been recorded (Johnson, Dingkuhn, Jones, & Mahamane, 1998). A 57% reduction in rice yield was documented with the *E. crus-galli* population of 9 plants per meter square (Maun & Barrett, 1986). Additionally, higher densities of *E. crus-galli* may remove up to 80% of the soil nitrogen, especially at vegetative growth stages (Holm et al., 1991). Therefore, a good weed management strategy relies on the depletion of weed seed bank in top 0 – 6 cm of soil.

## 1.3 The Microwave Option

Microwave energy, as an alternative weed management strategy because of herbicide resistance, is being evaluated during last decade at Dookie Campus, The University of Melbourne. All the previous experiments have yielded a profound effect of microwave energy on weed suppression in different scenarios (Brodie et al., 2015; Brodie, Hamilton, & Woodworth, 2007; Brodie et al., 2009; Brodie & Hollins, 2015). This book endeavours to capture the knowledge gained from this work. Some of the highlights include further understanding of microwave energy distribution into soil for pre-emergence weed suppression under field conditions, and its unexplored effects on abundance and behaviour of soil ammonia oxidizer bacterial and archaeal community, and their role to increase plant available nitrogen for better crop growth, which could be consider as supplementary effect of microwave in addition to weed suppression.

Ultimately the purpose of this book is to share knowledge gained and encourage others to explore the prospects of microwave weed management techniques.

## 1.4 Reference

Bagavathiannan, M. V., Norsworthy, J. K., Smith, K. L., & Neve, P. (2011). Modeling the evolution of barnyardgrass (Echinochloa crus-galli L.) resistance to glyphosate in cotton and its management implications. Paper presented at the Beltwide Cotton Conference, Atlanta, Georgia.

Batlla, D., & Benech-Arnold, R. L. (2007). Predicting changes in dormancy level in weed seed soil banks: Implications for weed management. Crop Protection/Weed Science in Time of Transition, 26(3), 189-197.

Bebawi, F. F., Cooper, A. P., Brodie, G. I., Madigan, B. A., Vitelli, J. S., Worsley, K. J., & Davis, K. M. (2007). Effect of microwave radiation on seed mortality of rubber vine (Cryptostegia grandiflora R.Br.), parthenium (Parthenium hysterophorous L.) and bellyache bush (Jatropha gossypiifolia L.). Plant Protection Quarterly, 22(4), 136-142.

Brodie, G., Grixti, M., Hollins, E., Cooper, A., Li, T., & Cole, M. (2015). Assessing the Impact of Microwave Treatment on Soil Microbial Populations. Global Journal of Agricultural Innovation, Research & Development, 2(1), 25-32.

Brodie, G., Hamilton, S., & Woodworth, J. (2007). An assessment of microwave soil pasteurization for killing seeds and weeds. Plant Protection Quarterly, 22(4), 143-149.

Brodie, G., Harris, G., Pasma, L., Travers, A., Leyson, D., Lancaster, C., & Woodworth, J. (2009). Microwave soil heating for controlling ryegrass seed germination. Transactions of the American Society of Agricultural and Biological Engineers, 52(1), 295-302.

Brodie, G., & Hollins, E. (2015). The Effect of Microwave Treatment on Ryegrass and Wild Radish Plants and Seeds. Global Journal of Agricultural Innovation, Research & Development, 2(1), 16-24.

Chong, K. Y., Corlett, R. T., Yeo, D. C. J., & Tan, H. T. W. (2011). Towards a global database of weed risk assessments: a test of transferability for the tropics. Biological Invasions, 13(7), 1571-1577. doi:10.1007/s10530-010-9914-5

DAFF. (2006). Weeds. Canberra: Australian Department of Agriculture, Fisheries and Forestry.

Davis, A. S., Dixon, P. M., & Liebman, M. (2003). Cropping system effects on giant foxtail (Setariafaberi) demography. 2. Retrospective perturbation analysis. Weed Science, 51, 930–939.

Gallandt, E. R. (2006). How can we target the weed seedbank? Weed Science, 54, 588-596.

Gourd, T. (2002). Controlling weeds using propane generated flame and steam treatments in crop and non croplands. Retrieved from Santa Cruz, CA, U. S. A.:

Holm, L. G., Plucknett, D. L., Pancho, J. V., & Herberger, J. P. (1991). The World's Worst Weeds: Distribution and Biology. Malabar, Florida, USA.: The University Press of Hawaii.

Johnson, D. E., Dingkuhn, M., Jones, M. P., & Mahamane, M. C. (1998). The influence of rice plant type on the effect of weed competition on O. sativa and O. glaber¬rima. Weed Research, 38, 207-216.

Maun, M. A., & Barrett, S. C. H. (1986). The biology of Canadian weeds. 77. Echinochloa crus-galli (L.) Beauv. . Canadian Journal Plant Science, 66, 739–759.

Norsworthy, J. K., Burgos, N. R., Scott, R. C., & Smith, K. L. (2007). Consultant perspectives on weed management needs in Arkansas rice. Weed Technology, 21, 832-839.

Oerke, E. C. (2006). Crop losses to pests. Journal of Agricultural Sciences, 144, 31–43.

Oerke, E. C., & Dehne, H. W. (2004 ). Safeguarding production losses in major crops and the role of crop protection. Crop Protection, 23, 275–285.

Pimentel, D. (1997). Pest management in agriculture. In D. Pimentel (Ed.), Techniques for Reducing Pesticide Use: Environmental and Economic Benefits (pp. 1–11). Chichester, UK: John Wiley & Sons.

Pimentel, D. (2005). Environmental and economic costs of the application of pesticides primarily in the United States. Environment, Development and Sustainability, 7, 229-252.

Rao, A. N., Johnson, D. E., Sivaprasad, B., Ladha, J. K., & Mortimer, A. M. (2007). Weed management in direct seeded rice. Advances in Agronomy, 93, 153-255.

Research, & Markets. (2017). Global Herbicide Market 2016-2022 By Product Type, Mode of Action, Crop Type, Regions and Vendors - Market Size, Demand Forecasts, Industry Trends and Updates, Supplier Market Shares - Research and Markets.

# 2 The Growing Threat to Herbicide Use

## 2.1 Introduction

Agriculture is fundamental to feeding a growing human population. It has two key economic characteristics: first, it produces goods that directly satisfy basic human needs; and second, it combines human effort with natural resources, such as land and water (Diao, Hazell, Resnick, & Thurlow, 2007) to deliver these goods. It was believed that, since natural resources were freely available, agriculture could grow independently of other economic activities (Diao et al., 2007); however, arable land area is a fixed resource so growth is constrained by available land (Diao et al., 2007). Historical evidence suggests that agriculture is the key to economic development. Gollin, Parente, and Rogerson (2002) show that once agriculture switches from traditional to modern technology, labour is released to the industrial sector and the economy grows at higher rates. Unfortunately, this means that agriculture's percentage contribution to national productivity declines due to dilution from growth in other sectors of the economy that would not be possible without an efficient agricultural sector. Agriculture's declining share in the economy sends a confusing signal to policy makers who conclude that agriculture is relatively unimportant and the falling real prices of agricultural commodities, due to transformational improvements in efficiency, sends the signal to investors that returns from agricultural investment are unattractive (Stringer, 2001).

Most modern agricultural systems can be characterized as highly simplified and disturbed habitats where deliberately imposed monocultures interact with a variety of other species (Keren, Menalled, Weaver, & Robison-Cox, 2015). Despite this, most agriculturalists are aware of the impact they have on the environment, and especially the soil. There has been a rapid increase in the use of reduced tillage for crop production in Australia. In 2001, more than 40 percent of the cropping area in Australia was under no-till (Chauhan, Gill, & Preston, 2006). The portion of no-till cropping in Australia has increased significantly since then. No-till agriculture depends mainly on chemical weed control (Chauhan, 2006).

Modern weed science was born out of the advances in chemistry that emerged out of the Second World War (Menalled et al., 2016). Herbicide based weed management has helped establish minimum till and zero till cropping systems (Yu, Cairns, & Powles, 2007). According to Menalled et al. (2016), the unifying principles driving herbicide-based weed management are: (1) the precautionary principle, where the most lethal treatment is generally adopted; (2) rapid and effective response to treatment; and (3) an expectation of complete eradication. This is further compounded by management recommendations which are based on single-species tactics (Keren et al., 2015). Unfortunately, as Menalled et al. (2016) point out, the success of this approach and the reluctance to conduct integrated and multidisciplinary research has resulted in strong selection pressure for herbicide resistant biotypes. It is over 60 years since Harper predicted the evolution of resistance to herbicides (Harper, 1956). Globally, there are 400 weed species that have developed resistance to herbicides and annually nine new weed biotypes are reported as being herbicide resistant (Heap, 2016). The development of resistance is an inevitable consequence of reliance on chemistry for weed control (Menalled et al., 2016).

## 2.2 A Brief History of Herbicides

Herbicides are chemical substances or cultured biological organisms that kill or suppress plant growth by affecting one or more of the processes that are vital to plant survival. Herbicides applied at high rates kill all plants. At low rate, some herbicides kill some plants without damaging other plants. Herbicides with such ability are said to be selective. Selective herbicides are commonly used in weed control.

Herbicide trials were initiated in 1929 by the arrival of the chlorate weed killers (Wheeler, 1962). Tests showed that sodium or calcium chlorate sprays were quite effective in killing weeds such as leafy spurge (*Euphorbia esula*), hoary cress (*Cardaria spp.*) and some of the perennial thistles (*Cirsium arvense*). However, because they were non-selective and quite persistent, chlorate weed killers also have the unfortunate effect of rendering the soil unsuitable for cropping for two or more years after the application at the high rates necessary to provide effective control of perennial weeds. Despite their limitations, chlorates became the first chemical weed killers to gain acceptance.

By the late 1930, selective sprays such as sulfuric acid and the recently introduced French discovery, Sinox (sodium dinitro-ortho-cresylate), were become increasingly popular for weed control (Hoyman, 1947) , but their use was strictly limited to heavy weed infestations or to areas where intensive land use was practiced because they were expensive. Consequently, they never seriously challenged sodium chlorate's position as the most commonly used chemical weed killer.

Chlorate weed killers are significant as a harbinger of the developments that followed, for generating sustained herbicide research. This was revealed in 1945 with the commercial release of 2,4-dichlorophenoxyacetic acid, commonly known as 2,4-D (Marcinkowska, Praczyk, Gawlak, Niemczak, & Pernak, 2017). Discovered independently in England and United States in the early 1940s, 2,4-D was the product of plant hormone research rather than the agriculturists' random search for phytotoxic compounds and it proved to be far more potent than any herbicide previously developed. Applications of approximately 140 g $ha^{-1}$ of the active ingredient was required to kill a wide range of broad-leaf species. More importantly, 2, 4-D is also highly selective, for when it is applied at these rates, most grasses are unaffected. The potential use for such chemical in land dominated by cereal production and awash in broad leaf weeds is obvious. Consequently, it was credited as a new era in farming. For many agriculturists, the arrival of 2, 4-D could not have been timelier, when science had suddenly delivered the ultimate anti-weed weapon just as farmers were on the verge of admitting defeat in their war for economic survival.

Between the years 1950 and 1970, the initial discovery and development of triazine herbicides took place (Montgomery & Freed, 1964). Their discovery and development were important scientific achievements that led to unprecedented success in crop weed management. Within a short time, the discovery of and screening for the herbicidal properties of dialkylamino-s-triazines led to selection of the most promising candidates for field development and eventual commercial use. For five decades, the triazines have provided weed control in more than 50 crops around the world.

Glyphosate, a phosphonomethyl derivative of glycine, was invented in 1950 as a pharmaceutical compound (Cox, 2004). Nevertheless, since the discovery of its herbicidal properties in 1970 and its commercialization in 1974 (Cox, 2004), glyphosate has been used in crop lands and non-crop lands. Glyphosate being non-selective was initially limited to pre-plant, post directed, and postharvest applications for weed control. With the introduction of glyphosate-resistant crops in the mid-1990s, glyphosate is now widely used for weed control. Over use of one herbicide has consequences.

## 2.3 The Development of Herbicide Resistance

Harper (1956) predicted the evolution of resistance to herbicides over 60 years ago. Globally, there are now 400 weed species that have developed resistance to various herbicides and annually 9 new weed biotypes are reported as being herbicide resistant (Heap, 2016). According to Neve (2007) between 2001–2005, 12% of papers published in the journal *Weed Research* reported studies on herbicide resistant; therefore, herbicide resistance has become a significant issue. With few exceptions, one or more of three general mechanisms confers herbicide resistance: an altered herbicide target enzyme; enhanced herbicide metabolism; or reduced herbicide translocation (Neve, 2007).

There have been three major stages in the development of herbicide resistance: resistance to Weed Science Society of America (WSSA) Group 5 (photosystem II inhibitor herbicides) was first reported in 1970; this was followed by WSSA Group 2 (acetolactate synthase [ALS] inhibitor herbicides) resistance in 1982; and in 1996 with WSSA Group 9 (enzyme 5-enolpyruvylshikimate 3-phosphate synthase-inhibitor herbicide) (Heap, 2016). For each of these herbicide groups, it took some additional time before a major agronomic impact in most crop systems was recognized (Owen, 2016), and most other herbicide groups now have weed biotypes with evolved resistance(s) in many economically important weed species (Heap, 2016). The development of WSSA Group 9 resistance was of major importance, given the unprecedented global adoption of glyphosate-resistant (GR) crop cultivars. Development of herbicide resistance in a population can be very quick. Field experiments, conducted by Hugh J. Beckie and Reboud (2009), demonstrated that almost 100% of the seed bank of field pennycress (*Thlaspi arvense*), growing in wheat crops, showed ALS inhibitor resistance within only four years.

Cross-resistance in weed flora is resistance to two or more herbicides of the same or different chemistry because of one resistant mechanism (RM) (Beckie & Tardif, 2012); however, multiple resistances in individual weed species is generally characterized by the presence of two or more RM. These mechanisms might be the mutation at the site of action (SOA) of herbicides (target site) or change in metabolism and translocation (non-target site), which reduces the phytotoxic effect of herbicides on their SOA (Beckie, Warwick, & A., 2012). Metabolic resistance is more commonly found in monocot (grasses) than in dicot (broadleaf) weeds (Beckie et al., 2012). Herbicide resistance in weeds is the greatest threat to sustainable productivity of agricultural commodities in industrialized countries. Therefore, there is a present need of an alternative weed management strategy in exiting cropping system. Unfortunatly, a combination of the cheapness of herbicide weed management and investment of significant infrastructure into no-till cropping systems has deterred the search for alternatives.

Shaner and Beckie (2014) provided a review of failures and mistakes of weed control approaches, which are centred on chemical solutions. Problem areas include: a general lack of diversity in weed management technologies; an unwillingness by weed scientists to conduct integrated and multidisciplinary research; a reluctance by extension specialists to provide complex recommendations to growers; a reluctance for growers to accept difficult-to-implement solutions; and the short-term priorities of herbicide marketing and sales organizations determining recommendations rather than considering longer-term strategies for herbicide stewardship (Keren et al., 2015; Menalled et al., 2016; Shaner & Beckie, 2014).

Another concern for weed management is a significant decline in herbicide discovery investment. Very little recent work has been done to develop new chemicals for weed management (Owen, 2016; Pucci, 2016). No new herbicides with novel sites of action have been commercially introduced in almost past three decades, and no new herbicide sites of action have been identified and developed (Owen, 2016). According to Stephen O. Duke (2012), there are probably several reasons for a lack of new herbicide development: new potential products may have remained dormant owing to concerns that glyphosate-resistant crops have reduced the market for a new herbicide; the capture of a large fraction of the herbicide market by glyphosate significantly diminished herbicide discovery efforts; company consolidations; and the availability of more generic herbicides. Another problem might be that the best herbicide molecular target sites may have already been discovered (Duke, 2012). Whatever the cause may be, no major new mode of action has been introduced to the market place for about 20 years (Duke, 2012). Before this, a new mode of action was introduced approximately every three years, leading to current use of approximately

20 known modes of action (Duke, 2012). Declining herbicide efficacy and lack of new development are being further exacerbated by social concerns about herbicide usage.

## 2.4 Indirect Costs of Herbicide Usage

Worldwide, about three Megatonnes of pesticides is applied to crops each year (Pimentel, 1995). Insecticides make up 20 to 30 percent of the total applied pesticide, about 50 to 60 percent are herbicides, and 10 to 20 percent are fungicides (Pimentel, 1995); therefore, up to two Megatonnes of herbicide, about 915,000 tonnes of active ingredient (Menalled et al., 2016), are applied to the planet's land mass every year. Despite the widespread application of pesticides at recommended dosage rates, pests (insects, plant pathogens, and weeds) still destroy about 37 percent of all potential crops (Pimentel, 2005). One reason for this production loss is that, in general, less than 0.1 percent of the applied chemical reaches the target pests (Pimentel, 1995). Herbicide treatments are probably more effective than this general figure for all pesticides; however, drift, off target applications, adverse weather conditions, entrainment of residual herbicides in rainfall runoff, poor adsorption, and poor translocation within the plant greatly reduce the herbicide efficacy.

Interest in chemical-free weed control has been increasing due to concerns over: herbicide resistance; the environmental; and human health impacts of herbicide use (Relyea, 2005; Wickerham et al., 2012). Recently, the International Agency for Research on Cancer (IARC), which is part of the World Health Organisation (WHO), has concluded that glyphosate is probably carcinogenic to humans (Guyton et al., 2015). This announcement has generated considerable debate in the media concerning the use of herbicides. Other authors have also highlighted the potential hazard to human health of exposure to herbicides and pesticides (Duke, 2010; Hernåndez et al.; Mačkić & Ahmetović, 2011; Peighambarzadeh, Safi, Shahtaheri, Javanbakht, & Forushani, 2011; Troudi et al., 2012; Wickerham et al., 2012).

Pimentel (2005) also points out that several other costs are incurred from pesticide application (Tab. 2.1). The annual indirect costs of herbicide application could be as much as US$5.8 billion in the United States and about US$433 million in Australia (Pimentel, 2005). Damage to crops may occur even when recommended dosages of herbicides are applied under normal environmental conditions (Pimentel, 1995, 2005). The increase in susceptibility of some crops to insects and disease damage, following the normal recommended use of 2,4-D and other herbicides has also been demonstrated (Pimentel, 1995).

**Table 2.1:** Total estimated environmental and social costs from pesticide in the United States.

| Costs | USA (population 318 Mil), Millions of US$/year | US$/year per person | Australia (population 23.8 Mil), Millions of US$/year |
|---|---|---|---|
| Public health impacts | 1140 | 3.58 | 85.2 |
| Domestic animal deaths and contaminations | 30 | 0.09 | 2.1 |
| Loss of natural enemies | 520 | 1.64 | 39.0 |
| Cost of pesticide resistance | 1500 | 4.72 | 112.3 |
| Honeybee and pollination losses | 334 | 1.05 | 25.0 |
| Crop losses | 1391 | 4.37 | 104.0 |
| Fishery losses | 100 | 0.31 | 7.38 |
| Bird losses | 2160 | 6.79 | 161.6 |
| Groundwater contamination | 2000 | 6.30 | 149.9 |
| Government regulations to prevent damage | 470 | 1.48 | 35.2 |
| Total | 9645 | 30.33 | 722 |

## 2.5 Conclusion

A combination of large scale industrialisation of agricultural production; the phenomenal efficacy and affordability of herbicides; the long term and significant investment in no-till infrastructure and benefit of no-till to soil health; rapid development of herbicide resistant biotypes; lack of investment in new discovery; a lack of integrated and multidisciplinary research; and the growing indirect costs and social concerns about wide spread herbicide usage and human health have brought modern agriculture to a point of convergence. While herbicides have served humanity very well for over 60 years, new ideas for weed management are needed.

## 2.6 References

Beckie, H. J., & Reboud, X. (2009). Selecting for Weed Resistance: Herbicide Rotation and Mixture. Weed Technology, 23(3), 363-370. doi:10.1614/wt-09-008.1

Beckie, H. J., & Tardif, F. J. (2012). Herbicide cross resistance in weeds. Crop Protection, 35, 15–28.

Beckie, H. J., Warwick, S. I., & A., S. C. (2012). Basis for herbicide resistance in Canadian populations of wild oat (Avena fatua). Weed Science, 60, 10–18.

Chauhan, B. S. (2006). Ecology and management of weeds under no-till in southern Australia. (PhD), University of South Australia, Roseworthy.

Chauhan, B. S., Gill, G., & Preston, C. (2006). Influence of tillage systems on vertical distribution, seedling recruitment and persistence of rigid ryegrass (Lolium rigidum) seed bank Weed Science, 54(4), 669-676.

Cox, C. (2004). Glyphosate. Journal of Pesticide Reform, 24(4), 10-15.

Diao, X., Hazell, P., Resnick, D., & Thurlow, J. (2007). The Role of Agriculture in Development: Implications for Sub-Saharan Africa. Retrieved from Washington, USA.
Duke, S. O. (2010). Herbicide and pharmaceutical relationships. Weed Science, 58(3), 334-339.
Duke, S. O. (2012). Why have no new herbicide modes of action appeared in recent years? Pest Management Science, 68(4), 505-512. doi:10.1002/ps.2333
Gollin, D., Parente, S., & Rogerson, R. (2002). The role of agriculture in development. American Economic Review, 92(2), 160-164.
Guyton, K. Z., Loomis, D., Grosse, Y., El Ghissassi, F., Benbrahim-Tallaa, L., Guha, N., et al. (2015). Carcinogenicity of tetrachlorvinphos, parathion, malathion, diazinon, and glyphosate. The Lancet Oncology, 16(5), 490-491. doi:10.1016/S1470-2045(15)70134-8
Harper, J. L. (1956). The evolution of weeds in relation to resistance to herbicides. Paper presented at the The 3rd British Weed Control Conference, Farnham, UK.
Heap, I. M. (2016). International Survey of Herbicide Resistant Weeds. Retrieved from http://www.weedscience.org/
Hernández, A. F., Parrón, T., Tsatsakis, A. M., Requena, M., Alarcón, R., & López-Guarnido, O. Toxic effects of pesticide mixtures at a molecular level: Their relevance to human health. Toxicology(0). doi:http://dx.doi.org/10.1016/j.tox.2012.06.009
Hoyman, W. M. (1947). Observations on the use of potato vine killers in the red river valley of North Dakota. American potato journal, 24(4), 110-116.
Keren, I. N., Menalled, F. D., Weaver, D. K., & Robison-Cox, J. F. (2015). Interacting Agricultural Pests and Their Effect on Crop Yield: Application of a Bayesian Decision Theory Approach to the Joint Management of Bromus tectorum and Cephus cinctus. Plos One, 10(2), 1-15. doi:10.1371/journal.pone.0118111
Mačkić, S., & Ahmetović, N. (2011). Toxicological profiles of highly hazardous herbicides with special reference to carcinogenicity to humans. Herbologia, 12(2), 55-60.
Marcinkowska, K., Praczyk, T., Gawlak, M., Niemczak, M., & Pernak, J. (2017). Efficacy of herbicidal ionic liquids and choline salt based on 2,4-D. Crop Protection, 98, 85-93. doi:http://doi.org/10.1016/j.cropro.2017.03.011
Menalled, F., Peterson, R., Smith, R., Curran, W., Páez, D., & Maxwell, B. (2016). The Eco-Evolutionary Imperative: Revisiting Weed Management in the Midst of an Herbicide Resistance Crisis. Sustainability, 8(12), 1297.
Montgomery, M. L., & Freed, V. H. (1964). Metabolism of Herbicides, Metabolism of Triazine Herbicides by Plants. Journal of Agricultural and Food Chemistry, 12(1), 11-14. doi:10.1021/jf60131a004
Neve, P. (2007). Challenges for herbicide resistance evolution and management: 50 years after Harper. Weed Research, 47(5), 365-369. doi:10.1111/j.1365-3180.2007.00581.x
Owen, M. D. K. (2016). Diverse Approaches to Herbicide-Resistant Weed Management. Weed Science, 64(sp1), 570-584. doi:10.1614/WS-D-15-00117.1
Peighambarzadeh, S. Z., Safi, S., Shahtaheri, S. J., Javanbakht, M., & Forushani, A. R. (2011). Presence of Atrazine in the Biological Samples of Cattle and Its Consequence Adversity in Human Health. Iranian Journal of Public Health, 40(4), 112-121.
Pimentel, D. (1995). Amounts of pesticides reaching target pests: Environmental impacts and ethics. Journal of Agricultural and Environmental Ethics, 8(1), 17-29. doi:10.1007/bf02286399
Pimentel, D. (2005). Environmental and economic costs of the application of pesticides primarily in the United States. Environment, Development and Sustainability, 7, 229-252.
Pucci, J. (2016). 'Back to the Future' Weed Control: Herbicide Outlook 2016. CropLife. Retrieved from http://www.croplife.com/crop-inputs/herbicides/back-to-the-future-weed-control-herbicide-outlook-2016/

Relyea, R. A. (2005). The Lethal Impact of Roundup on Aquatic and Terrestrial Amphibians. Ecological Applications, 15(4), 1118-1124.

Shaner, D. L., & Beckie, H. J. (2014). The future for weed control and technology. Pest Management Science, 70(1329–1339).

Stringer, R. (2001). How important are the 'non-traditional' economic roles of agriculture in development? Retrieved from Adelaide:

Troudi, A., Sefi, M., Ben Amara, I., Soudani, N., Hakim, A., Zeghal, K. M., et al. (2012). Oxidative damage in bone and erythrocytes of suckling rats exposed to 2,4-dichlorophenoxyacetic acid. Pesticide Biochemistry and Physiology, 104(1), 19-27. doi:http://dx.doi.org/10.1016/j.pestbp.2012.06.005

Wheeler, A. F. J. (1962). Some experiments in total weed control. Journal of the Science of Food and Agriculture, 13(2), 103-108. doi:10.1002/jsfa.2740130208

Wickerham, E. L., Lozoff, B., Shao, J., Kaciroti, N., Xia, Y., & Meeker, J. D. (2012). Reduced birth weight in relation to pesticide mixtures detected in cord blood of full-term infants. Environment International, 47(0), 80-85. doi:http://dx.doi.org/10.1016/j.envint.2012.06.007

Yu, Q., Cairns, A., & Powles, S. (2007). Glyphosate, paraquat and ACCase multiple herbicide resistance evolved in a Lolium rigidum biotype. Planta, 225(2), 499-513.

# 3 A System Model for Crop Yield Potential as a Function of Herbicide Weed Control over Time

## 3.1 Introduction

Modelling is an all-encompassing term, which suggests anything from building a scaled representation of the system through to developing a rigorous mathematical analysis of the system's internal relationships. The ultimate purpose of the model is to produce a representation of the system (Wilson, 1988), which is easily understood, behaves in a similar way to the real system and can be more easily manipulated than the real system, in order to understand how things may change as inputs or time vary, which is referred to as "sensitivity analysis".

Often the system model is represented by mathematical relationships. There are two main strategies for determining these mathematical relationships. These are the empirical approach, which usually requires a statistical analysis of data from two or more system parameters (Walpole & Myers, 1972) to evaluate the degree of cause and effect between these parameters, or the deterministic analysis, which employs a range of clearly defined and rigorous mathematical procedures to define the relationship between two or more system parameters.

In terms of mathematical effort, empirical relationships are generally easier to develop, but can not be stretched beyond the sampling range of the original data (Walpole & Myers, 1972) from which the relationship was inferred. On the other hand, deterministic models are much harder to derive, requiring a sound knowledge of mathematic disciplines such as vectors, algebra and calculus, but the resulting equations tend to be much more robust than empirical relationships. Deterministic models can be extrapolated to explore more extreme cases.

System analysis can be applied to most agricultural systems to better understand their operation and optimise performance. System analysis usually includes the development of transfer functions (Åström & Murray, 2012; Smith, 1976). Transfer functions are mathematical equations, involving various input variables or matrices, which relate the system's output to these system inputs. In the case of an agricultural cropping system the key output from the system is potential crop yield. Some crop ecology studies have demonstrated that competition from weeds can reduce the potential yield of some crops by 35% to 55% (Cathcart & Swanton, 2003; Mondani, Golzardi, Ahmadvand, Ghorbani, & Moradi, 2011).

Modern no-till cropping depends on herbicides for weed management; therefore, herbicide applications are an important system input. Unfortunately, herbicide resistance in many weed species is also becoming wide spread (Heap, 1997) and multiple herbicide resistances in several economically important weed species has been widely reported (Owen, Walsh, Llewellyn, & Powles, 2007). In time, herbicide resistant weeds may result in significant yield reductions and grain contamination;

therefore, this chapter derives a deterministic system transfer function, relating herbicide input to potential crop yield, in the presence of herbicide resistance, based on various ecological models published in literature.

## 3.2 Derivation of Crop System Transfer Function for Herbicide Weed Management

The effect of weed damage on crop yields can be described by equation (3.1) (Schmidt & Pannell, 1996). The nomenclature of some parameters used in this chapter is outlined in Table 3.1, which is at the end of the chapter.

$$Y = Y_o\left[1 - D(R)\right] \tag{3.1}$$

In equation (3.1), D(R) is the damage function caused by a weed density of R, which represents the number of weeds that are recruited from the seed bank (plants $m^{-1}$ of row). The Damage function can be described by the following equation (Cousens, Brain, O'Donovan, & O'Sullivan, 1987):

$$D(R) = \frac{I \cdot R}{100\left(e^{c} + \frac{I \cdot R}{A_w}\right)} \tag{3.2}$$

Substituting equation (3.1) into equation (3.2) yields:

$$Y = Y_o\left[1 - \frac{I \cdot R}{100\left(e^{c} + \frac{I \cdot R}{A_w}\right)}\right] \tag{3.3}$$

## 3.3 Herbicide Weed Management

Weed infestations will be made up of some resistant weeds ($R_R$) and some weeds that can be easily controlled sby herbicides ($R_S$), where the total weed population is the sum of these two components (i.e. $R_R$ - (1 - $S$) and $R_S = S$). A typical kill function for a herbicide treatment is (Bosnić & Swanton, 1997):

$$K(H) = e^{-\lambda H} \tag{3.4}$$

Substituting all this into equation (3.4), and realising that the herbicide treatment will not affect the resistant weeds, yields:

$$Y = Y_o \left\{ 1 - \frac{I \cdot R_O \cdot \left[1 - S + S \cdot e^{-\lambda H}\right]}{100\left[e^{ct} + \frac{I \cdot R_O \cdot \left[1 - S + S \cdot e^{-\lambda H}\right]}{A_W}\right]} \right\} \tag{3.5}$$

The recruitment of seedlings from the seed bank can be described by the following equation (Neve, Norsworthy, Smith, & Zelaya, 2011):

$$R_o = \frac{W}{\left[1 + e^{-\left(\frac{t - t_O}{d}\right)}\right]} \tag{3.6}$$

Substituting this into equation (3.5) yields:

$$Y = Y_o \left\{ 1 - \frac{I \cdot W \cdot \left[1 - S + S \cdot e^{-\lambda H}\right]}{100\left[e^{ct}\left[1 + e^{-\left(\frac{t - t_O}{d}\right)}\right] + \frac{I \cdot W \cdot \left[1 - S + S \cdot e^{-\lambda H}\right]}{A_W}\right]} \right\} \tag{3.7}$$

The portion of the population that is resistant to herbicide treatment will change from generation to generation depending on the selection pressure being applied by the herbicide treatments. Based on work by Gubbins and Gilligan (1999), if there is a relatively constant selection pressure (a) towards herbicide resistance from generation to generation, then the following relationship will hold:

$$\frac{\partial S}{\partial g} = -aSg \tag{3.8}$$

This partial differential equation can be solved by integration to give:

$$S = S_o \cdot e^{\frac{-ag^2}{2}} \tag{3.9}$$

Substituting this into equation (3.7) and simplifying yields:

$$Y = Y_o \left\{ 1 - \frac{I \cdot W \cdot \left[1 - S_O \cdot e^{\frac{-ag^2}{2}} + S_O \cdot e^{\frac{-ag^2}{2} - \lambda H}\right]}{100\left[e^{ct}\left[1 + e^{-\left(\frac{t - t_O}{d}\right)}\right] + \frac{I \cdot W \cdot \left[1 - S_O \cdot e^{\frac{-ag^2}{2}} + S_O \cdot e^{\frac{-ag^2}{2} - \lambda H}\right]}{A_W}\right]} \right\} \tag{3.10}$$

There is also evidence that herbicides have a toxic effect on the crop as well. Using the study by Yin et al. (2008) as a guide, and assuming that the toxicity of the herbicide on a crop can be expressed as a polynomial of the form $Loss=aH^2 - bH$, equation (3.10) can be modified to become:

$$Y = Y_o \left\{ 1 - \frac{I \cdot W \cdot \left[1 - S_o \cdot e^{\frac{-ag^2}{2}} + S_o \cdot e^{\frac{-ag^2}{2} - \lambda H}\right]}{100 \left[ e^{ct}\left[1 + e^{-\left(\frac{t-t_o}{d}\right)}\right] + \frac{I \cdot W \cdot \left[1 - S_o \cdot e^{\frac{-ag^2}{2}} + S_o \cdot e^{\frac{-ag^2}{2} - \lambda H}\right]}{A_W} \right]} + aH^2 - bH \right\} \tag{3.11}$$

The seed bank will be dynamic depending on factors such as natural seed mortality, immigration of seeds into the area from other locations via various vectors, emigration of seeds out of the area to other locations via various vectors, the onset of dormancy that prevents germination in the current season, and the breaking of dormancy from previous seasons in the seed bank.

$$Y = Y_o \left\{ 1 - \frac{I \cdot [W(1-N-D_o) - E_m + I_m] \cdot \left[1 - S_o \cdot e^{\frac{-ag^2}{2}} + S_o \cdot e^{\frac{-ag^2}{2} - \lambda H}\right]}{100 \left\{ e^{ct}\left[1 + e^{-\left(\frac{t-t_o}{d}\right)}\right] + \frac{I \cdot [W(1-N-D_o) - E_m + I_m] \cdot \left[1 - S_o \cdot e^{\frac{-ag^2}{2}} + S_o \cdot e^{\frac{-ag^2}{2} - \lambda H}\right]}{A_W} \right\}} + aH^2 - bH \right\} \tag{3.12}$$

## 3.4 Sensitivity Analysis

The development of transfer functions does not always provide accurate prediction but to provides insight into system behaviours as input parameters change. The sensitivity of the output to these changes can be assessed by differentiating the transfer function equations with respect to the input parameter of interest and assessing the magnitude of the resulting differential equation. For example, the sensitivity of the crop to herbicide weed control is given by differentiating equation (3.12) with respect to the herbicide dose, H:

$$\frac{\partial Y}{\partial H} = Y_o \left\{ \frac{I \cdot [W(1-N-D_o) - E_m + I_m] \cdot e^{ct}\left[1 + e^{-\left(\frac{t-t_o}{d}\right)}\right] \cdot A_W{}^2 \cdot \lambda \cdot S_o \cdot e^{\frac{-ag^2}{2} - \lambda H}}{100 \left\{ e^{ct}\left[1 + e^{-\left(\frac{t-t_o}{d}\right)}\right] A_W + I \cdot [W(1-N-D_o) - E_m + I_m] \cdot \left[1 - S_o \cdot e^{\frac{-ag^2}{2}} + S_o \cdot e^{\frac{-ag^2}{2} - \lambda H}\right] \right\}^2} + 2aH - b \right\} \tag{3.13}$$

Herbicide resistance in many weed species is becoming more prevalent (Heap, 1997, 2008). Thornby and Walker (2009) simulated continuous summer fallows using glyphosate. Their modelling showed that barnyard grass (*Echinochloa colona*) could become resistant to glyphosate in about 15 years. Validation of their model against paddock history data for glyphosate-resistant population of barnyard grass showed that their model correctly predicted resistance development to within a few years of the real situation.

Selection pressure for genetic traits depends on the initial efficacy of the herbicide to remove susceptible individuals from the population, leaving only the resistant individuals to reproduce. This is reinforced by the adoption of a single herbicide over a long period to sustain the selection pressure on the population.

The transfer function developed in equation (3.12) can also provide some insight in the rate of change of yield potential as a function of weed population generations, hence providing some insights into herbicide resistance. Differentiating equation (3.12) with respect to the generations of weeds gives:

$$\frac{\partial Y}{\partial g} = Y_o \left\{ \frac{I \cdot [W(1-N-D_o)-E_m+I_m] \cdot e^{ct}\left[1+e^{-\left(\frac{t-t_o}{d}\right)}\right] \cdot A_w{}^2 \cdot S_o \cdot a \cdot g \cdot \left(e^{\frac{-ag^2}{2}} - e^{\frac{-ag^2}{2}-\lambda H}\right)}{\left(100\left\{e^{ct}\left[1+e^{-\left(\frac{t-t_o}{d}\right)}\right]A_w + I \cdot [W(1-N-D_o)-E_m+I_m] \cdot \left[1-S_o \cdot e^{\frac{-ag^2}{2}} + S_o \cdot e^{\frac{-ag^2}{2}-\lambda H}\right]\right\}^2\right)} \right\} \tag{3.14}$$

Timeliness of herbicide application is another important consideration in weed management. Herbicide application can be delayed for several reasons, but often it is associated with inclement weather conditions such as wind and rain, both of which impede the opportunity to spray herbicides safely and effectively. If weeds become well established before the crop canopy closes, yield losses can be expected. The sensitivity of yield potential to timeliness can be evaluated by differentiating equation (3.12), with respect to t:

$$\frac{dY}{dt} = Y_o \left\{ \frac{I \cdot W \cdot \left[1 - S_o \cdot e^{\frac{-ag^2}{2}} + S_o \cdot e^{\frac{-ag^2}{2}-\lambda H}\right] \cdot c \cdot e^{ct}}{100\left[1 + e^{-\left(\frac{t-t_o}{d}\right)} + \frac{I \cdot W \cdot \left[1 - S_o \cdot e^{\frac{-ag^2}{2}} + S_o \cdot e^{\frac{-ag^2}{2}-\lambda H}\right]}{A_w}\right]} \right.$$

$$\left. - \frac{I \cdot W \cdot \left[1 - S_o \cdot e^{\frac{-ag^2}{2}} + S_o \cdot e^{\frac{-ag^2}{2}-\lambda H}\right] \cdot e^{ct}\left[1 + e^{-\left(\frac{t-t_o}{d}\right)}\right]}{100 \cdot d \cdot \left[1 + e^{-\left(\frac{t-t_o}{d}\right)} + \frac{I \cdot W \cdot \left[1 - S_o \cdot e^{\frac{-ag^2}{2}} + S_o \cdot e^{\frac{-ag^2}{2}-\lambda H}\right]}{A_w}\right]^2} \right\} \tag{3.15}$$

## 3.5 Examples

Equation (3.12) was coded into a simple cropping system model using the MatLab® software platform. Using data published by Bosnić and Swanton (1997) and Yin et al. (2008) for Rimsulfuron herbicide and assuming: an initially same small resistant population (i.e. $S_0$ =0.9999); a seed mortality rate of 10% each year; and a slightly positive selection coefficient of (a = 0.002) for herbicide resistance (Baucom & Mauricio, 2004), the system transfer function was used to analyse the effect of a single herbicide application on crop yield potential. The transfer function was also used to forecast the long-term crop yield potential, if only a single herbicide type was used during this time.

Figure 3.1 shows the expected crop yield response as a function of the herbicide's application rate. Based on the parameters used in this example, there is an optimal active ingredient application rate (i.e. where $\frac{\partial Y}{\partial H} = 0$) of about 0.009 kg ha$^{-1}$, while the maximum rate of crop yield response occurs at about 0.001 kg ha$^{-1}$.

The transfer function also predicts that significant herbicide resistance will occur within 15 generations (Figure 3.2), as was also predicted by Thornby and Walker (2009). This is apparent when looking at how the relative crop yield potential reduces along the generations axis in Figure 3.2. After 15 to 20 years of using the same herbicide control system, the model outlined in equation (3.12) suggest that further herbicide application will be ineffectual. Herbicide rotations can forestall the development of a resistant population; however several weed species have developed multiple resistance to several herbicide groups (Owen et al., 2007).

It is possible to visualise the influence of both herbicide application and generational change in a response surface, as shown in Figure 3.3.

The sensitivity of crop yield potential to timeliness can be assessed from equations (3.12) and (3.15). Figure 3.4 depicts the influence of time between crop emergence and weed emergence over crop yield potential.

## 3.6 Conclusion

A growing herbicide resistance problem is already evident in most Australian cropping systems (Broster & Pratley, 2006; Gill & Holmes, 1997). There is evidence that glyphosate resistance has already developed in some weed populations (Broster & Pratley, 2006) and multiple herbicide resistances has been widely reported in several weed species (Kuk, Burgos, & Talbert, 2000; Owen et al., 2007; Walsh, Powles, Beard, Parkin, & Porter, 2004; Yu, Cairns, & Powles, 2007); therefore significant crop yield losses can be expected into the future as weed become more resistant to herbicide management strategies. Alternative weed management strategies that are compatible with no-till cropping systems need to be developed. The next chapter will discuss some of the non-chemical weed control strategies that have been considered in recent time.

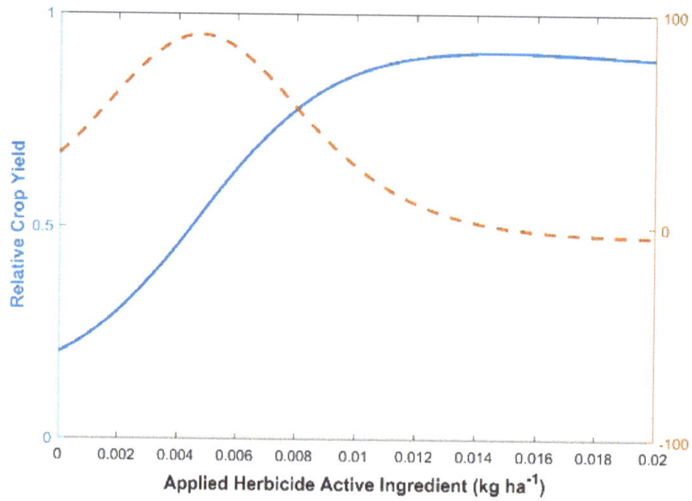


**Figure 3.1:** Normalised crop yield (blue line) and rate of change of crop yield (orange line) as a function of applied herbicide energy, based equation (3.12).

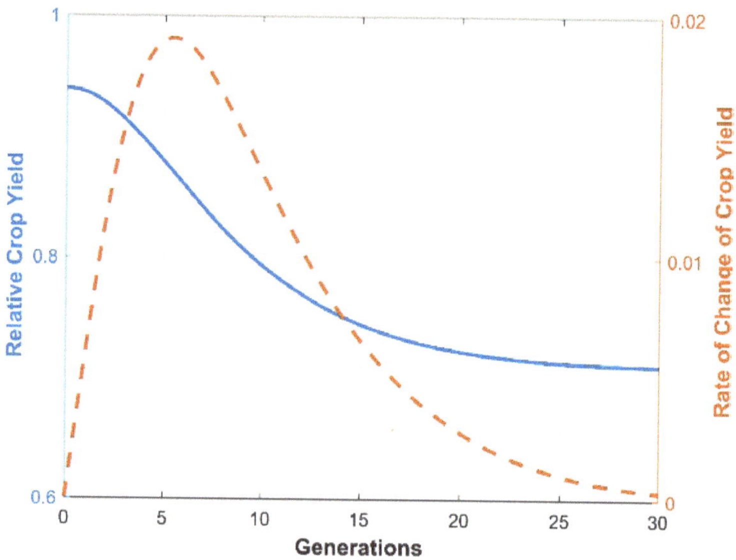


**Figure 3.2:** Normalised crop yield (blue line) and rate of change of crop yield (orange line) as a function of time (generations of weeds), based on (3.12).

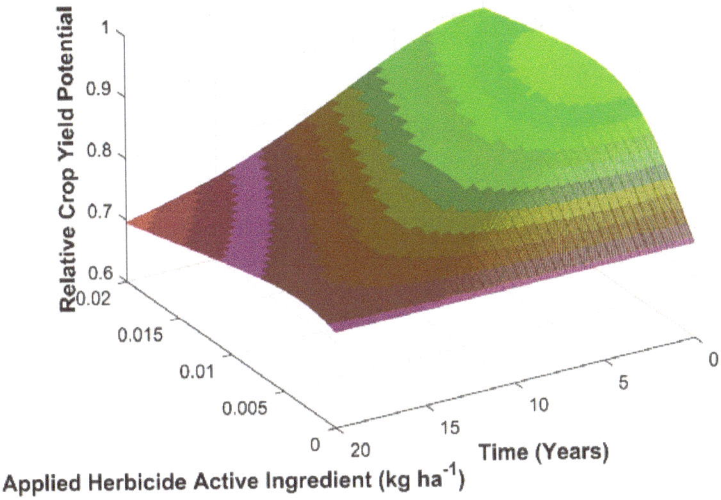


**Figure 3.3:** Response surface for potential crop yield as a function of both herbicide application and generational change in the weed population.

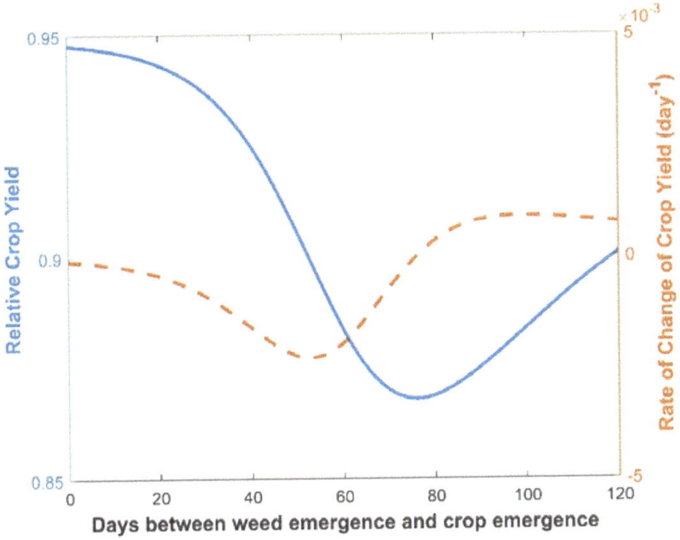


**Figure 3.4:** Response of crop yield potential to the number of days between weed emergence and crop emergence.

**Table 3.1:** Nomenclature used in this chapter.

| | |
|---|---|
| a | Is the selection pressure for herbicide resistance |
| $A_w$ | Is the percentage yield loss as weed density approaches ∞ (= 38.0 (Bosnić & Swanton, 1997)) |
| c | Is the speed of light (m $s^{-1}$) or the rate at which I approaches zero as t approaches ∞ (= 0.017 (Bosnić & Swanton, 1997)) |
| d | Is the slope of the seed bank recruitment curve at $t_0$ |
| $D_b$ | Fraction of the seed population from previous seasons breaking dormancy (Note: this is expressed as a fraction of the initial seed bank population $W_0$) |
| $D_0$ | Fraction of the seed population developing dormancy (Note: this is expressed as a fraction of the initial seed bank population $W_0$) |
| $E_m$ | Seed emigration from the area of interest |
| g | Is the generational number |
| H | Is the herbicide's active ingredient dose (kg $ha^{-1}$) |
| I | Is the percentage yield loss as the weed density tends towards zero (= 0.38 (Bosnić & Swanton, 1997)) |
| Im | Seed immigration into the area of interest |
| N | Is the natural death rate for the whole population (Note: this is expressed as a fraction of the initial seed bank population $W_0$) |
| $S_0$ | Is the initial frequency of plants in the population that are susceptible to herbicide treatment |
| $S_s$ | Viable seed set per plant from surviving volunteers in the weed population |
| t | Is the time difference between crop emergence and weed emergence |
| $t_0$ | Is the time for 50% germination of the viable seed bank |
| W | Is the viable seed bank |
| $Y_0$ | Is the theoretical yield with no weed infestations |
| λ | Is an estimate of weed sensitivity to the herbicide |

## 3.7 References

Åström, K. J., & Murray, R. M. (2012). Feedback Systems: An Introduction for Scientists and Engineers. Princeton, New Jersey: Princeton University Press.

Baucom, R. S., & Mauricio, R. (2004). Fitness costs and benefits of novel herbicide tolerance in a noxious weed. Proceedings of the National Academy of Sciences of the United States of America, 101(36), 13386-13390. doi:10.1073/pnas.0404306101

Bosnić, A. Č., & Swanton, C. J. (1997). Economic Decision Rules for Postemergence Herbicide Control of Barnyardgrass (Echinochloa crus-galli) in Corn (Zea mays). Weed Science, 45(4), 557-563.

Broster, J. C., & Pratley, J. E. (2006). A decade of monitoring herbicide resistance in Lolium rigidum in Australia. Australian Journal of Experimental Agriculture, 46(9), 1151-1160.

Cathcart, R. J., & Swanton, C. J. (2003). Nitrogen management will influence threshold values of green foxtail (Setaria viridis) in corn. Weed Science, 51(6), 975-986. doi:10.1614/p2002-145

Cousens, R., Brain, P., O'Donovan, J. T., & O'Sullivan, P. A. (1987). The use of biologically realistic equations to describe the effects of weed density and relative time of emergence on crop yield. Weed science (USA).

Gill, G. S., & Holmes, J. E. (1997). Efficacy of cultural control methods for combating herbicide-resistant Lolium rigidum. Pesticide Science, 51(3), 352-358.

Gubbins, S., & Gilligan, C. A. (1999). Invasion Thresholds for Fungicide Resistance: Deterministic and Stochastic Analyses. Proceedings: Biological Sciences, 266(1437), 2539-2549. doi:10.2307/1353823

Heap, I. M. (1997). The occurrence of herbicide-resistant weeds worldwide. Pesticide Science, 51(3), 235-243.

Heap, I. M. (2008). International Survey of Herbicide Resistant Weeds. Retrieved from http://www.weedscience.org/in.asp

Kuk, Y. I., Burgos, N. R., & Talbert, R. E. (2000). Cross- and multiple resistance of diclofop-resistant Lolium spp. Weed Science, 48(4), 412-419.

Mondani, F., Golzardi, F., Ahmadvand, G., Ghorbani, R., & Moradi, R. (2011). Influence of Weed Competition on Potato Growth, Production and Radiation Use Efficiency. Notulae Scientia Biologicae, 3(3), 42-52.

Neve, P., Norsworthy, J. K., Smith, K. L., & Zelaya, I. A. (2011). Modelling evolution and management of glyphosate resistance in Amaranthus palmeri. Weed Research, 51(2), 99-112.

Owen, M., Walsh, M., Llewellyn, R., & Powles, S. (2007). Widespread occurrence of multiple herbicide resistance in Western Australian annual ryegrass (Lolium rigidum) populations. Australian Journal of Agricultural Research, 58(7), 711-718.

Schmidt, C. P., & Pannell, D. J. (1996). Economic Issues in Management of Herbicide-Resistant Weeds. Research in Agricultural and Applied Economics, 64(3), 301-308.

Smith, R. J. (1976). Circuits, Divices and Systems (3rd ed.). New York: Wiley International.

Thornby, D. F., & Walker, S. R. (2009). Simulating the evolution of glyphosate resistance in grains farming in northern Australia. Annals Of Botany, 104(4), 747-756. doi:10.1093/aob/mcp152

Walpole, R. E., & Myers, R. H. (1972). Probability and Statistics for Engineers and Scientists. New York: Macmillan Publishing Co.

Walsh, M. J., Powles, S. B., Beard, B. R., Parkin, B. T., & Porter, S. A. (2004). Multiple-Herbicide Resistance across Four Modes of Action in Wild Radish (Raphanus raphanistrum). Weed Science, 52(1), 8-13.

Wilson, J. (1988). Changing agriculture: an introduction to systems thinking. Kenthurst [N.S.W.]: Kangaroo Press.

Yin, X. L., Jiang, L., Song, N. H., & Yang, H. (2008). Toxic Reactivity of Wheat (Triticum aestivum) Plants to Herbicide Isoproturon. Journal of Agricultural and Food Chemistry, 56(12), 4825-4831. doi:10.1021/jf800795v

Yu, Q., Cairns, A., & Powles, S. (2007). Glyphosate, paraquat and ACCase multiple herbicide resistance evolved in a Lolium rigidum biotype. Planta, 225(2), 499-513.

# 4 Physical Weed Control

## 4.1 Introduction

Modern no-till cropping depends on herbicides for weed management; therefore, herbicide applications are an important system input. Various estimates of the embodied energy needed to manufacture and distribute herbicides have been developed. Fore, Porter, and Lazarus (2011) suggest that the total energy associated with herbicide application is between 1.4 and 1.5 GJ $ha^{-1}$, depending on the crop they were assessing.

Unfortunately, herbicide resistance in many weed species is becoming wide spread (Heap, 1997, 2016) and multiple herbicide resistances in several economically important weed species has also been widely reported (Owen, Walsh, Llewellyn, & Powles, 2007). In time, herbicide resistant weeds may ultimately result in significant yield reductions and grain contamination.

The International Agency for Research on Cancer (IARC), which is part of the World Health Organisation (WHO), has also concluded that glyphosate is probably carcinogenic to humans (Guyton et al., 2015). This announcement has generated considerable debate in the media concerning the use of herbicides. Other authors have also highlighted the potential hazard to human health of long term exposure to herbicides and pesticides (Duke, 2010; Hernández et al., 2013; Mačkić & Ahmetović, 2011; Peighambarzadeh, Safi, Shahtaheri, Javanbakht, & Forushani, 2011; Troudi et al., 2012; Wickerham et al., 2012); therefore, there has been growing interest in non-herbicidal control of weeds. The objectives of this chapter are to outlines some of the potential technologies, apart from herbicide application, for weed management. These technologies include: flaming; steam treatment; electrocution; applying electrostatic fields; microwave weed treatment; applying infra-red radiation; applying ultraviolet radiation; using lasers; robotics; and using abrasive weed control techniques.

## 4.2 Tillage

Prior to the introduction of herbicides in the 1940's, weed control was often achieved through tillage (Price & Kelton, 2011). Tillage techniques include: hand hoeing; scarifying; or ploughing. Tillage physically disrupts plants, preventing them from maturing and setting seeds. Shallow tillage can stimulate seed bank germination, breaking dormancy in some seeds through abrasion of the seed coat and therefore allowing a follow up treatment to better control the emerged weeds. Tillage can also bury seeds deep enough in the soil profile that they do not emerge after germination.

### 4.2.1 Hand Hoeing

Hand hoeing of weeds has probably been undertaken since agriculture began. Giampietro and Pimentel (1990) assume that adult males can sustain a power output of 90 W, while adult women can sustain a 60 W power output. They based this assessment on ergonomic studies and a consistent 30% difference in elite athletic performance between men and women. The average human power output is therefore approximately 75 W.

Amery, Schramm, and Shapiro (1978) present data indicating that hand weed management in sorghum crops, in Central Niger, requires between 400 and 1000 man-hours annually. It has also been shown that humans require approximately 100 J of food energy to produce 1.0 J of work. Therefore, the energy required to hand hoe one hectare of land, assuming 400 man-hours for weed control, will be:

$$E = 400 \times 75 \times 3600 \times 100 = 10.8 \text{ GJ ha}^{-1} \quad (4.1)$$

Based on the same analysis, Giampietro and Pimentel (1991), demonstrate that using animals and machines reduce the required energy (Tab. 4.1).

**Table 4.1:** Comparison of energy requirements for different weed tillage control systems.

| Power Source | Gross-energy Requirements (GJ ha$^{-1}$) |
|---|---|
| Manpower | 10.8 |
| Oxen | 5.2 |
| 6-HP tractor | 3.1 |
| 50-HP tractor | 4.1 |

Source: Giampietro & Pimentel, 1991

### 4.2.2 Mechanical Tillage

Tillage force, required to draw a plough through the soil, depends on the width of the tine, the penetration depth into the soil, the plough design (i.e. whether it is a chisel plough, disc plough, or mould board plough) the travel speed (Saunders, Godwin, & O'Dogherty, 2000), the bulk density of the soil and soil internal shearing resistance (Godwin, O'Dogherty, Saunders, & Balafoutis, 2007; Saunders et al., 2000). Plough draught force increases as all of these parameters increase (Godwin et al., 2007; Saunders et al., 2000).

Traction is associated with the interface between a tyre and the soil and it profoundly influences the energy requirements for tillage. The soil exerts a force on the wheel, which is in response to the torque applied to the drive wheel by the vehicle's transmission system and engine (Zoz & Grisso, 2003). This response, or reaction force, is called the traction force.

As would be expected, the interface between the soil and the drive wheels does not perfectly transfer the tractor's motive force to ploughs. Travel reduction (Zoz & Grisso, 2003), which has traditionally been called "wheel slip" occurs between surfaces. Travel reduction occurs because of:

- Flexing of the drive wheels
- Slip between the surfaces (rubber and concrete, for example)
- Shear within the soil.

From a power efficiency standpoint, travel reduction represents a power loss caused by a loss in travel speed or distance (Zoz & Grisso, 2003). In practice, there is always travel loss, but it becomes significant when the vehicle is towing heavy loads or being used for draught work, such as pulling a plough. Traction efficiency is significantly reduced as travel reduction ratio increases (Figure 4.1). The relationship between travel reduction ratio and traction efficiency is described by:

$$\mathrm{TE} = \frac{a \cdot TRR^3 + b \cdot TRR^2 + c \cdot TRR + d}{TRR + \mathrm{e}} \tag{4.2}$$

Tabatabaeefar, Emamzadeh, Varnamkhasti, Rahimizadeh, and Karimi (2009) performed an assessment of several tillage systems to determine the input energy needed (Tab. 4.2).

Intense soil disturbance often leads to soil degradation, erosion and loss of productivity (Price & Kelton, 2011); therefore, modern agriculture is focused on reducing tillage. Therefore, weed management strategies need to be effective, but without engaging the soil.

## 4.3 Thermal Weed Control

Thermal weed control (flaming and steam) applies heat directly to the weed, which quickly raises the temperature of the moisture in the plants cambium cells. The rapid expansion of this moisture causes the cell structure to rupture, preventing nutrients and water from entering the stalk and leaves (Gourd, 2002).

Thermodynamics predicts that energy, in the form of heat, moves along the temperature gradient until all spatial coordinates reach equilibrium. Equilibrium is reached when the temperature gradient disappears from the system.

Heat is transferred by conduction, convection and radiation. Conduction is the transfer of heat between solid/solid interfaces, and within solids. Convection is the

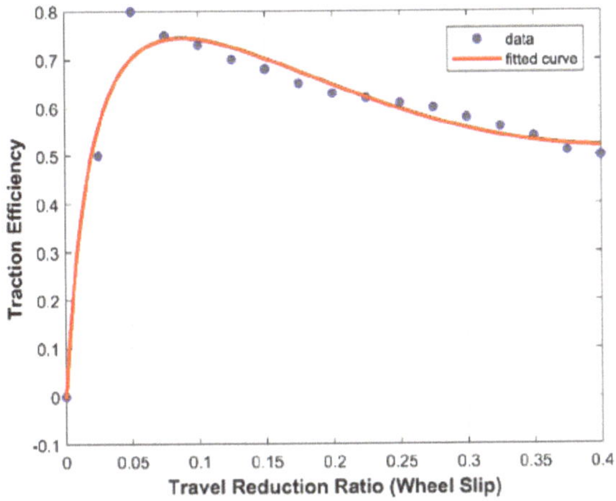


**Figure 4.1:** Relationship between Travel Reduction Ratio and Traction Efficiency for a MF 6270 Massy-Ferguson tractor (tested at the Dookie Campus of Melbourne University).

**Table 4.2:** Comparison of energy requirements for various forms of mechanical tillage (Source: Tabatabaeefar et al., 2009).

| Treatment | Input Energy (GJ ha$^{-1}$) |
|---|---|
| Mouldboard plough | 18.71 |
| Chisel plough | 17.62 |
| Cyclo-tiller | 17.67 |
| No-till | 16.33 |

transfer of heat between an object and its environment due to fluid motion, i.e. a gas or liquid interface with a solid. Radiation is the transfer of heat between bodies through the emission and absorption of electromagnetic energy, without the need for a fluid interface (i.e. as a purely spatial phenomenon). This section will explore how heat can be used to kill weeds.

### 4.3.1 Flaming

Flame weeding is the most commonly applied thermal weed control method. Several kinds of equipment have been developed for weeding, such as tractor-mounted flamers and hand-pushed or handheld devices for weeding around obstacles and for private households. Flaming controls a wide range of weed species (Ascard, 1994), some of which are tolerant or resistant towards herbicides. Flaming gave 72 and 80% control of common rye and volunteer alfalfa, respectively. Both kochia (*Kochia scoparia* (L.)

Roth) and netseed lambsquarter (*Chenopodium berlandieri*) were also controlled at 65% (Gourd, 2002).

Ascard (1994) developed dose response relationships between applied energy and weed response. He used three models to determine responses; however, his results are mostly based on the model presented in equation (4.3).

$$y = \frac{D}{1+\left(\frac{x}{a}\right)^{b}} \tag{4.3}$$

where y is the response variable of the plant fresh weight or plant number, x is the liquid petroleum gas (LPG) consumption in kg $ha^{-1}$, and D, a and b are parameters to be determined experimentally. From this model, an $LD_{50}$ and $LD_{95}$ were derived for white mustard (*Sinapis alba* L.) at different plant sizes and densities (Tab. 4.3).

Weed flamers should be shielded, preferably with a long and relatively low roofed shield (Storeheier, 1994) to keep combustion gases close to the ground for as long as possible; the burner angle should be 22.5° to 45° to the horizontal.

Tandem burners did not increase effective ground speed compared with single burners (Ascard, 1998). According to Ascard (1994), propane doses of 10–40 kg $ha^{-1}$ were required to achieve 95% control of sensitive species with 0–4 leaves, whilst plants with 4–12 leaves required 40–150 kg $ha^{-1}$. At 49 MJ $kg^{-1}$, this corresponds to 7.35 GJ $ha^{-1}$ or 73.5 J $cm^{-2}$.

Species with protected meristems, such as Shepherd's purse (*Capsella bursa-pastoris* L.), were tolerant due to regrowth after flaming and they could only be completely killed in their early stages. Annual bluegrass (*Poa annua* L.) could not be completely killed with a single flame treatment, regardless of developmental stage or propane dose. Considerably lower doses (40%) were required in years with higher precipitation compared with a dry year. Because precipitation enhances thermal weed control efficacy, a system which induces high humidity could provide better weed control.

### 4.3.2 Steam Treatment

Steam based weed control has received renewed interest in recent years. The most common and simplest steam applicator is sheet steaming. This involves covering the soil with a thermally resistant membrane, which is sealed at the edges. Steam, which is pumped under the sheet, penetrates the soil's surface layer to kill weeds and their seeds (Gay, Piccarolo, Ricauda Aimonino, & Tortia, 2010b). A more mobile option is to use a small hooded applicator head, connected to a steam source via a hose, to apply saturated steam to the soil surface (Gay et al., 2010b).

Gay et al. (2010b) tested a hooded applicator with an area of 150 mm by 150 mm. Their steam generator had a nominal duty cycle of 8.5 kW with a 1.6 kW superheater

**Table 4.3:** Parameter estimates of regression model for plant number data after flame treatment of white mustard at different plant sizes and densities (Modified from: Ascard, 1994).

| Number of leaves | Plant density (No. $m^{-2}$) | D from equation (4.18) (No. $m^{-2}$) | a = $LD_{50}$ (kg $ha^{-1}$) | b from equation (4.18) | $LD_{95}$ (kg $ha^{-1}$) | $LD_{95}$ (GJ $ha^{-1}$) |
|---|---|---|---|---|---|---|
| 0-2 | 195 | 174 | 21.7 | 4.55 | 41.5 | 2.03 |
| 0-2 | 395 | 342 | 21.7 | 4.55 | 41.5 | 2.03 |
| 2-4 | 169 | 155 | 38.8 | 4.55 | 74.1 | 3.63 |
| 2-4 | 365 | 335 | 38.8 | 4.55 | 74.1 | 3.63 |
| 0-2 | 250 | 207 | 22.1 | 3.87 | 47.3 | 2.31 |
| 0-2 | 714 | 658 | 22.1 | 3.87 | 47.3 | 2.31 |
| 3-4 | 265 | 210 | 35.6 | 4.76 | 66.1 | 3.24 |
| 3-4 | 798 | 607 | 60.4 | 3.01 | 159.5 | 7.82 |

to deliver 4 kg $h^{-1}$ of superheated steam. It heated an area of soil (100 mm by 50 mm) to a temperature of 90°C or more in 200 seconds. After 300 seconds the entire soil surface under the applicator reached a uniform temperature of 100°C (Gay et al., 2010b). Given that the thermal capacity of water is 4.2 kJ $kg^{-1}$ $°C^{-1}$ and the latent heat of vaporisation for water is 2.26 MJ $kg^{-1}$, and assuming an initial water temperature of 25°C, this represents an application energy of 13.5 kJ $cm^{-2}$.

Raffaelli et al. (2016) developed a band steaming system for field work. In their investigation of the system's performance, they used the following relationship for weed survival as a function of steam application:

$$Y = \frac{D-C}{1+e^{\{b[log(x)-log(LD_{50})]\}}} + C \tag{4.4}$$

Where Y is the weed response (plants $m^{-2}$), D is the upper limit of response (plants $m^{-2}$), C is the lower limit of response (plants $m^{-2}$), x is the applied steam (kg $m^{-2}$), and $LD_{50}$ is the applied steam needed to achieve a 50% weed mortality rate (kg $m^{-2}$).

The parameter values from their experimental trials with the system were: b = 2.6; C = -2.8; D = 72.5; and $LD_{50}$ = 1.0 (kg $m^{-2}$) (Raffaelli et al., 2016). Based on these values, the $LD_{90}$ (90% weed control) for this system was 2.3 kg $m^{-2}$ of steam.

Fennimore and Goodhue (2016) report on an experiment using a commercial steam soil treatment system, which treated 240 $m^2$ $h^{-1}$. The system requires 2.6 GJ $h^{-1}$, which implies an energy requirement of 108.3 GJ $ha^{-1}$ or 1,083 J $cm^{-2}$. The energy expended during steam treatment varies considerably (Tab. 4.4). Gourd (2002) reports on an experiment where 125 US gallons of steam (water) was used and 3600 square feet (334.4 $m^2$) of weeds. Based on the thermal properties of water, this equates to 365.6 J $cm^{-2}$.

Kolberg and Wiles (2002) reported on their experiments using steam to treat emerged weeds. They discovered that a treatment 890 kJ $m^{-2}$ was necessary to achieve

**Table 4.4:** Reported energies used in steam treatments.

| Authors | Treatment being applied | Applied Energy Density (J cm$^{-2}$) |
|---|---|---|
| Nishimura, Asai, Shibuya, Kurokawa, and Nakamura (2015) | Soil fumigation | 55,430 |
| Gay et al. (2010b) | Soil fumigation | 13,500 |
| Fennimore and Goodhue (2016) | Soil fumigation | 1,083 |
| Gelsomino, Petrovičová, Zaffina, and Peruzzi (2010) | Soil fumigation but with CaO addition for extra thermal activity | 781 |
| Gourd (2002) | Weed treatment | 366 |
| Melander and Kristensen (2011) | Soil fumigation but with band steaming only | 325 |
| Raffaelli et al. (2016) | Soil fumigation but with band steaming only | 199 |
| Kolberg and Wiles (2002) | Weed treatment | 89 |
| Rask et al. (2013) | Weed treatment (on hard surfaces) | 75 |

a similar level of weed control to glyphosate; however, in several cases their steam treatment was not completely effective. Rask, Larsen, Anderson, and Kristofferson (2013) demonstrated that 41.3 GJ ha$^{-1}$ year$^{-1}$ was required to control weeds on hard surfaces such as foot paths, traffic islands, and path ways. Their observations suggest that 5.5 treatments were needed annually for effective weed control; therefore, the energy required for a single treatment was approximately 7.5 GJ ha$^{-1}$, or 75 J cm$^{-2}$.

Soil pasteurisation can also be achieved by injecting steam into the soil using gridded steam injectors (Gay, Piccarolo, Ricauda Aimonino, & Tortia, 2010a; Gay et al., 2010b). This technique could be used as an alternative to soil fumigation, which is commonly applied in high value horticultural crops.

Gay et al. (2010b) developed a scalar index to measure heating efficiency for steam soil heating

$$I = \frac{1}{Vt_f}\int_0^{t_f}\int_V T \cdot dV \cdot dt \tag{4.5}$$

Where V is the volume of soil being heated (m$^3$), T is the temperature increase (K), and $t_f$ is the heating time (s). Gay et al. (2010b) applied this index to their steam experiments. Equation (4.5) represents the 4-D average of the temperature change in the soil volume. The performance of sheet steaming varied between about 7.5 and 18, the hooded applicator varied between about 27 and 37 and the steam injection system varied between 37 and 47 (Gay et al., 2010b).

**Table 4.5:** Summary of non-chemical weed control data from Rask et al. (2013).

| Treatment Technology | Mean number of treatments per year for effective weed management | Mean energy requirements (GJ ha$^{-1}$) |
|---|---|---|
| Flame | 5 | 34.6 |
| Hot air/flame | 5.5 | 67.9 |
| Steam | 5.5 | 41.3 |
| Hot water | 3 | 43.1 |

While steam treatment is effective at killing weeds, and can achieve some pasteurisation of the soil (Gay et al., 2010a, 2010b), it requires considerable energy investment to create the steam. This is partly due to the inherent limitations of convective heat transfer.

### 4.3.3 Hot Water

Hot water treatment is somewhat linked to steam treatment; however, the initial energy input is less, because there is no need to incur the latent heat requirements of 2.27 MJ kg$^{-1}$, associated with turning liquid water into steam. In spite of this, some studies have shown that hot water alone is insufficient to provide good weed control and various methods of holding the heat in the plants need to be used. These methods include: multiple applications of hot water; the draping of some kind of thermal blanket behind the applicator, or application of insulating foam along with the hot water (Kempenaar & Spijker, 2004).

In an assessment of non-chemical weed control for use on hard surfaces, Rask et al. (2013) determined the mean dose of propane gas per year, needed to control weeds, based on different technologies. These are summarised in terms of applied energy per hectare in Table 4.5. From this data it is apparent that hot water is not significantly different from steam treatment, in terms of applied energy.

## 4.4 Radiation Systems

Radiation can be used to overwhelm weeds with energy. Ultimately, radiation based weed control is a form of heat treatment.

### 4.4.1 Infrared Radiation

Heat kills plants, there being a time-temperature relationship (Levitt, 1980). Radiative heat transfer refers to the transfer of energy by broad spectrum electromagnetic radiation from some adjacent hot object (or from a hot environment) to the heated object. Any object that is above zero degrees Kelvin will radiate energy in the form of electromagnetic photons. The German physicist, Max Planck (1858 – 1947), deduced that the radiation spectral density (ρ) given off from a hot object depended on the wavelength of interest and the temperature of the object. This spectral density can be described by:

$$\rho = \frac{2hc^2}{\lambda^5\left\{e^{\frac{hc}{\lambda kT}} - 1\right\}} \tag{4.6}$$

Where h is Planck's constant ($6.6256 \times 10^{-34}$ J s), c is the speed of light, λ is the electromagnetic wavelength of interest, k is Boltzmann's constant ($1.38054 \times 10^{-23}$ J $K^{-1}$), and T is the temperature in Kelvin. A typical set of spectral distributions for different temperatures is shown in Figure 4.2.

The brightness temperature of a body can be determined by rearranging Planck's equation to find T for a given spectral density value:

$$T = \frac{hc^2}{\lambda k \cdot \ln\left(\frac{2\pi hc}{\rho\lambda^5} + 1\right)} \tag{4.7}$$

The wavelength at which peak radiation intensity occurs can be found by differentiating Planck's equation and setting the derivative equal to zero (Appendix A). Therefore, the wavelength of peak radiation is determined by:

$$\lambda_p \approx \frac{hc}{5kT} \tag{4.8}$$

Where $\lambda_p$ is the peak radiation wave length (m). At room temperature, or above, the wavelength of peak radiation will be in the micrometre range (~10 μm), which is in the Long-wavelength Infrared Band (Table 4.6). The penetration of electromagnetic energy into materials is limited by the wavelength and the dielectric properties of the material (Vollmer, 2004):

$$\delta = \frac{\lambda_p}{4\pi\sqrt{\kappa}} \tag{4.9}$$

Where: δ is the penetration depth (m) and κ is the relative dielectric constant of the material. The penetration depth of any radiation from objects at room temperature

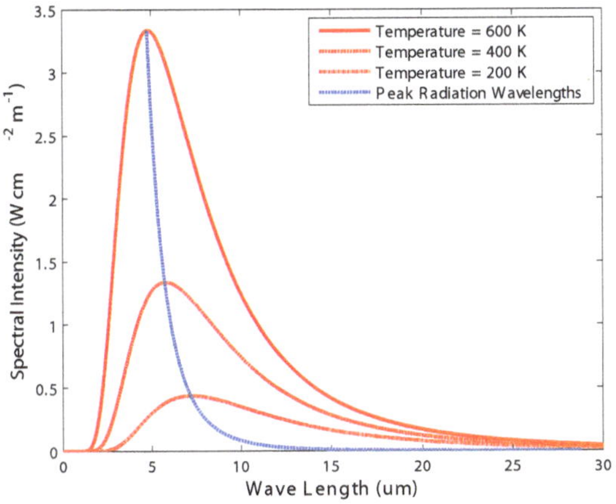


**Figure 4.2:** Radiative spectral density at different temperatures as a function of temperature and wavelength.

**Table 4.6:** A commonly used sub-division scheme.

| Division Name | Abbreviation | Wavelength (μm) | Temperature (K) |
|---|---|---|---|
| Near Infrared | NIR | 0.75 – 1.4 | 3,964 – 2,070 |
| Short-wavelength Infrared | SWIR | 1.4 – 3.0 | 2,070 – 966 |
| Mid-wavelength Infrared | MWIR | 3.0 – 8.0 | 966 – 362 |
| Long-wavelength Infrared | LWIR | 8.0 – 15.0 | 362 – 193 |
| Far Infrared | FIR | 15.0 – 1,000 | 193 - 3 |

or above will be in the nanometre range; therefore, radiative heat transfer must be regarded as a surface phenomenon where further heat transfer from the surface into the material occurs via internal conduction and convection.

The total radiated power can be determined by integrating Planck's equation across all wavelengths for a particular temperature (Appendix B) to yield the Stefan-Boltzmann equation. The power transferred from an object at one temperature to another object at a lower temperature is given by (Holman, 1997):

$$q = \varepsilon\sigma A\left(T_A^4 - T_p^4\right) \tag{4.10}$$

Where q is the radiation power (W); ε is the surface emissivity of the radiator material; σ is the Stefan-Boltzmann constant ($5.6704 \times 10^{-8}$ J $s^{-1}$ $m^{-2}$ $K^{-4}$); A is the surface area of the heated object ($m^2$); $T_A$ is the temperature of the infrared applicator (K); and $T_p$ is the temperature of the plants being treated (K).

Denaturing of plant cell components starts with long term exposure to temperatures of about 40°C. The fatal impacts of high temperatures on plants have been studied in detail for over a century (Levitt, 1980). In particular, a thoroughly demonstrated empirical relationship between lethal temperature and temperature holding time has been developed by Lepeschkin (1912):

$$T = 79.8 - 12.8 \cdot log_{10}Z \tag{4.11}$$

Where T is the lethal temperature (°C), and Z is the lethal temperature holding time, in minutes (Levitt, 1980).

Infrared radiation systems use gas burners to heat ceramic or metal surfaces, which then radiate infrared energy towards the ground. According to Parish (1990), laboratory investigations identified that a '*medium wave tubular fused quartz infrared emitter*' was the most effective for weed control. Infrared burners are not affected by wind, in contrast to flame weeders, and they cover a more closely defined area.

Ascard (1998) discovered that efficacy of flaming and infrared radiation treatment, on emerging seedlings, was similar. For example, when white mustard (*Sinapis alba* L.) plants were at the 4-leaf stage, propane doses of 8 kg $ha^{-1}$ from either flaming or infrared systems merely scorched the edges of the leaves. Propane doses of 30 kg $ha^{-1}$ desiccated almost 20% of the plants, but surviving plants showed vigorous re-growth. One hundred percent weed control required 120 kg $ha^{-1}$ of propane for both systems.

Considerably higher temperatures were required under the flamer compared to the infrared radiator. Temperatures of up to 1,350°C were recorded in the central blue part of the stationary flamer system; however, the stationary infrared radiator had a maximum temperature of 770°C (Ascard, 1998). The ground temperature in both cases was approximately 180°C (Ascard, 1998). Ascard (1998) also reports work by Hoffmann who found that infrared radiators cause a higher temperature increase in the upper few millimetres of soil compared to flamers, because radiation heating avoids the convective heat transfer limitations, which are associated with hot air (flame) heating.

For efficient plant destruction, an infrared radiator is required, which produces high energy intensity at a wavelength which is absorbed, rather than reflected or transmitted, by the plant tissues. To kill young white mustard plants, an energy density at ground level of between 200 kJ $m^{-2}$ and 400 kJ $m^{-2}$ (or 20 to 40 J $cm^{-2}$) of short wave or medium wave infrared energy are required to severely restrict plant growth (Parish, 1990). These dose rates are similar to those associated with microwave weed control discussed in the previous section; however, because microwaves have a much longer wave length than infrared radiation, the penetration of microwave energy into plants and the soil will be much further.

In several studies, infrared radiators have proved to be inferior compared with flame weeders, but Ascard (1998) and Parish (1990) found the differences in effect was dependent on the type of thermal weeder, dose, ground speed, burner height, plant size, plant density and plant species. Their studies also indicate that infrared burners are more likely to suffer from shading interference in dense vegetation compared with flame weeders that cause turbulence and thereby expose more leaves to the flame. This shading effect is linked to the shallow penetration of infrared radiation into most dielectric materials. Because microwave energy has a much longer wave length than infrared energy, microwave weed control is less vulnerable to shading, than other radiation systems, including ultraviolet radiation.

### 4.4.2 Ultraviolet Radiation

The wavelength of ultraviolet (UV) radiation lies between 100 and 400 nm and is thus outside the visible range. UV rays can be separated into three groups on the basis of wave length: UV-A (320 to 400 nm), UV-B (280 to 320 nm), and UV-C (100 to 280 nm). When plants are irradiated with UV, almost all energy is absorbed in the outermost 0.1- to 0.2-mm layer of the plant tissue. This results in heating of the plant tissue and thus can have effects similar to the damage to plants from flame weeding (Andreasen, Leif, & Jens, 1999).

Andreasen et al. (1999) irradiated four weed species at two different leaf stages and two crops at one leaf stage with ultraviolet light from a water cooled 2.35 kW UV lamp. The weed species were: annual bluegrass (*Poa annua* L.); common groundsel (*Senecio vulgaris* L.); shepherd's purse (*Capsella bursa-pastoris* (L.) Medicus); and small nettle (*Urtica urens* L.). The crop species were: canola (*Brassica napus* L. ssp. napus); and pea (*Pisum sativa* L.). Plants were treated in a laboratory with the UV lamp placed as close as possible to the plant canopy without touching it (about 1 cm above). Plant parts close to the lamp received more radiation than parts farther away. After irradiation, the above-ground fresh weight was measured after the plants were withered, but before re-growth commenced from undamaged buds.

Andreasen et al. (1999) used the following model to fit their data:

$$Y = \frac{D-C}{1+e^{\{b[log(x)-log(LD_{50})]\}}} + C \tag{4.12}$$

Where Y is the fresh weight yield to a UV dose of x (GJ $ha^{-1}$). D is the upper limit of fresh weight and C is the lower limit (g $pot^{-1}$), and b and $LD_{50}$ are determined experimentally. Table 4.7 lists the dose response parameters from their experiemnt. Smaller plants are more susceptible to ultraviolet radiation than larger plants, as indicated by their $LD_{50}$'s, which are somewhat similar in magnitude to both infrared and microwave radiation, described earlier.

Andreasen et al. (1999) observed re-growth after irradiation, which suggests that more than one treatment would be necessary to obtain efficient weed control. They also deiscovered that the distance between the source of UV radiation and the target plants played an important role: increasing the distance from just above the canopy to 17 cm increased the required dose almost two-fold (Andreasen et al., 1999).

### 4.4.3 Lasers

Light Amplification through Stimulated Emission of Radiation (Laser) is commonly used for cutting industrial materials, surgery, wood cutting, and for research. Laser creates coherent, monochromatic light, which concentrates a large amount of energy into a narrow, non-spreading beam (Heisel, Schou, Christensen, & Anderson, 2001). Recently, UV (355 nm), visible (532 nm), IR (810 nm), and $CO_2$ (1064 nm) lasers have been used to cut the stems of weeds, including perennial ryegrass (*Lolium perenne* L.) (Heisel et al., 2001).

Mathiassen, Bak, Christensen, and Kudsk (2006) investigated the effect of laser treatment on common chickweed (*Stellaria media*), scentless mayweed (*Tripleurospermum inodorum*), and canola (*Brassica napus*). Effective treatment requires the laser to be focused onto the apical meristem of the plants (Mathiassen et al., 2006). Several machine vision based systems have been explored to achieve accurate placement of the laser spot onto weed plants (Mathiassen et al., 2006). Another technique is to move the laser beam back and forth as the system moves forward to achieve good ground coverage. In all cases, it is essential that the laser beam intercepts the weed plant in a favourable way that causes damage to the stem. Proper laser focusing is difficult to achieve in practical terms.

Heisel et al. (2001) found that applying between 0.9 and 2.3 J mm$^{-2}$ from a $CO_2$ laser, when applied below the meristem, resulted in a 90% or more reduction in weed biomass, in common lamb's quarters (*Chenopodium album*) and wild mustard (*Sinapis arvensis*), respectively. Mathiassen et al. (2006) developed a response equation for laser weed treatment of the form:

$$Y = \frac{D-C}{1+e^{\left[2b\left(\log(LD_{90})+\frac{1.099}{b}-\log(x)\right)\right]}} + C \tag{4.13}$$

Where Y is the fresh weight yield to a UV dose of x (J mm$^{-2}$), D is the upper limit of fresh weight, C is the lower limit, and b and $LD_{90}$ are determined experimentally. Their results indicate that the efficacy of laser weed control depends on the weed species, wavelength, exposure time, spot size and laser power (Tab. 4.8). As with all radiation weed control methods, efficacy increases with power and exposure time; however, in the case of laser based weed control, spot diameter also affects efficacy. The most efficient system was the 5 W, 532 nm laser with a 1.8 mm spot diameter (Mathiassen et al., 2006).

**Table 4.7:** Summary of regression parameters from the estimated dose-response curves (Modified from: Andreasen et al., 1999).

| | Species | Growth stage | D (g pot$^{-1}$) | C (g pot$^{-1}$) | b | $LD_{50}$ (GJ ha$^{-1}$) | $LD_{50}$ (J cm$^{-2}$) |
|---|---|---|---|---|---|---|---|
| Weeds | Annual | I | 2.51 | -- | 0.98 | 1.23 | 12.3 |
| | bluegrass | II | 11.8 | -- | 0.74 | 11.1 | 111.0 |
| | Common | I | 3.32 | 0.023 | 1.38 | 0.50 | 5.0 |
| | groundsel | II | 17.1 | -- | 0.76 | 6.22 | 62.2 |
| | Small nettle | I | 2.56 | 0.03 | 2.02 | 0.10 | 1.0 |
| | | II | 16.4 | 6.92 | 1.34 | 1.48 | 14.8 |
| | Shepherd's | I | 3.18 | -- | 0.67 | 0.16 | 1.6 |
| | purse | II | 8.94 | 0.86 | 0.97 | 0.5 | 5.0 |
| Crops | Canola | I | 20.7 | 0.25 | 1.24 | 0.75 | 7.5 |
| | Pea | I | 5.45 | 0.45 | 1.07 | 3.13 | 31.3 |

**Table 4.8:** Plant response to laser treatment (Modified from: Mathiassen et al., 2006).

| Laser | Spot Diameter (mm) | Stellaria media | | | Tripleurospermum inodorum | | | Brassica napus | | |
|---|---|---|---|---|---|---|---|---|---|---|
| | | b | $LD_{90}$ (J) | $LD_{90}$ (J mm$^{-2}$) | b | $LD_{90}$ (J) | $LD_{90}$ (J mm$^{-2}$) | b | $LD_{90}$ (J) | $LD_{90}$ |
| 5 W, | 0.9 | -4.6 | 1.4 | 2.2 | -3.4 | 2.6 | 4.1 | n.e. | > 5.0 | 7.8 |
| 532 nm | 1.8 | n.e. | < 1.25 | 0.5 | -5.3 | 2.7 | 1.1 | -4.7 | 10.8 | 4.3 |
| 90 W, | 1.2 | -3 | 58.3 | 51.6 | -5.4 | 44.8 | 39.7 | n.e. | > 90.0 | 79.6 |
| 810 nm | 2.4 | -0.9 | 104.9 | 23.2 | -3.2 | 73.8 | 16.3 | -1.6 | > 225.0 | 49.8 |

**Note:** n.e. indictaes that this paramater was unable to be evaluated.

Most of the values for b in Table 4.8 are large, indicating that the slope of the plant response is high; therefore, it is important to apply an energy dose which is higher than the threshold value for $LD_{90}$, to ensure efficacy. When the various laser energy doses are scaled for comparison to other weed control techniques, a dose of between 50 and 7960 J cm$^{-2}$ of laser energy is required to control 90% of weeds, depending on the weed species, growth stage, laser power, and laser wavelength, which is more than is needed by some other forms of electromagnetic radiation.

Lasers have the potential to provide weed control; however, the device needs to accurately target the weed plant stems to kill the plant. This is not easily achieved. Various machine vision or scanning techniques are being investigated to provide accurate laser targeting for weed control. These systems are also being used in autonomous agricultural robots.

**Table 4.9:** Summary of energy needs for various weed control strategies on a per treatment basis.

| Weed or Soil Treatment System | Mean energy Requirements for complete coverage (GJ $ha^{-1}$) |
|---|---|
| Steam – soil fumigation | 1,190 |
| Laser | 233.5 |
| Steam – Weed control | 17.7 |
| Tillage | 17.5 |
| Hot water | 14.3 |
| Hand Hoeing | 10.8 |
| Ox drawn tillage | 5.2 |
| Flaming | 3.4 |
| IR | 3 |
| UV | 2.5 |
| Herbicide (Total energy, including embodied energy for manufacture and transport) | 1.4 |

## 4.5 Conclusions

All the techniques discussed in this chapter can be used to either control weed growth or kill weeds. Most of the technologies discussed in this chapter require moderate to high energy investment to achieve adequate weed control (Tab. 4.9). Some of these technologies have been commercialised to some degree; however, many of them have not progressed beyond the research phase. As herbicide resistance becomes more prevalent, some, or all, of these technologies may become more widely adopted.

## 4.6 References

Amery, D., Schramm, G., & Shapiro, K. (1978). Production Systems in Fragile Environments. Paper presented at the Science and Technology for Managing Fragile Environments in Developing Nations.

Andreasen, C., Leif, H., & Jens, C. S. (1999). The Effect of Ultraviolet Radiation on the Fresh Weight of Some Weeds and Crops. Weed Technology, 13(3), 554-560.

Ascard, J. (1994). Dose–response models for flame weeding in relation to plant size and density. Weed Research, 34(5), 377-385. doi:10.1111/j.1365-3180.1994.tb02007.x

Ascard, J. (1998). Comparison of flaming and infrared radiation techniques for termal weed control. Weed Research, 38, 69-76.

Duke, S. O. (2010). Herbicide and pharmaceutical relationships. Weed Science, 58(3), 334-339.

Fennimore, S. A., & Goodhue, R. E. (2016). Soil Disinfestation with Steam: A Review of Economics, Engineering, and Soil Pest Control in California Strawberry. International Journal of Fruit Science, 16(S1), 71-83.

Fore, S. R., Porter, P., & Lazarus, W. (2011). Net energy balance of small-scale on-farm biodiesel production from canola and soybean. Biomass and Bioenergy, 35(5), 2234-2244. doi:10.1016/j.biombioe.2011.02.037

Gay, P., Piccarolo, P., Ricauda Aimonino, D., & Tortia, C. (2010a). A high efficacy steam soil disinfestation system, part II: Design and testing. Biosystems Engineering, 107(3), 194-201. doi:http://dx.doi.org/10.1016/j.biosystemseng.2010.07.008

Gay, P., Piccarolo, P., Ricauda Aimonino, D., & Tortia, C. (2010b). A high efficiency steam soil disinfestation system, part I: Physical background and steam supply optimisation. Biosystems Engineering, 107(2), 74-85. doi:http://dx.doi.org/10.1016/j.biosystemseng.2010.07.003

Gelsomino, A., Petrovičová, B., Zaffina, F., & Peruzzi, A. (2010). Chemical and microbial properties in a greenhouse loamy soil after steam disinfestation alone or combined with CaO addition. Soil Biology and Biochemistry, 42(7), 1091-1100. doi:http://dx.doi.org/10.1016/j.soilbio.2010.03.006

Giampietro, M., & Pimentel, D. (1990). Assessment of the energetics of human labor. Agriculture, Ecosystems & Environment, 32(3), 257-272. doi:https://doi.org/10.1016/0167-8809(90)90164-9

Giampietro, M., & Pimentel, D. (1991). Energy efficiency: Assessing the interaction between humans and their environment. Ecological Economics, 4(2), 117-144. doi:https://doi.org/10.1016/0921-8009(91)90025-A

Godwin, R. J., O'Dogherty, M. J., Saunders, C., & Balafoutis, A. T. (2007). A force prediction model for mouldboard ploughs incorporating the effects of soil characteristic properties, plough geometric factors and ploughing speed. Biosystems Engineering, 97(1), 117-129. doi:10.1016/j.biosystemseng.2007.02.001

Gourd, T. (2002). Controlling weeds using propane generated flame and steam treatments in crop and non croplands. Retrieved from Santa Cruz, CA, U. S. A.

Guyton, K. Z., Loomis, D., Grosse, Y., El Ghissassi, F., Benbrahim-Tallaa, L., Guha, N., et al. (2015). Carcinogenicity of tetrachlorvinphos, parathion, malathion, diazinon, and glyphosate. The Lancet Oncology, 16(5), 490-491. doi:10.1016/S1470-2045(15)70134-8

Heap, I. M. (1997). The occurrence of herbicide-resistant weeds worldwide. Pesticide Science, 51(3), 235-243.

Heap, I. M. (2016). International Survey of Herbicide Resistant Weeds. Retrieved from http://www.weedscience.org/

Heisel, T., Schou, J., Christensen, S., & Anderson, C. (2001). Cutting weeds with a CO2 laser. Weed Research, 41, 19-29.

Hernández, A. F., Parrón, T., Tsatsakis, A. M., Requena, M., Alarcón, R., & López-Guarnido, O. (2013). Toxic effects of pesticide mixtures at a molecular level: Their relevance to human health. Toxicology, 134-145. doi:http://dx.doi.org/10.1016/j.tox.2012.06.009

Holman, J. P. (1997). Heat Transfer (8th ed.). New York: McGraw-Hill.

Kempenaar, C., & Spijker, J. H. (2004). Weed control on hard surfaces in The Netherlands. Pest Management Science, 60(6), 595-599. doi:10.1002/ps.863

Kolberg, R. L., & Wiles, L. J. (2002). Effect of Steam Application on Cropland Weeds. Weed Technology, 16(1), 43-49. doi:10.1614/0890-037X(2002)016[0043:EOSAOC]2.0.CO;2

Lepeschkin, W. W. (1912). Zur Kenntnis der Einwirkung supamaximaler Temperaturen auf die Pflanze. Berichte der Deutschen Botanishen Gesellschaft, 30, 713-714.

Levitt, J. (1980). Response of Plants to Environmental Stresses (Vol. 1). New York: Academic Press.

Mačkić, S., & Ahmetović, N. (2011). TOXICOLOGICAL PROFILES OF HIGHLY HAZARDOUS HERBICIDES WITH SPECIAL REFERENCE TO CARCINOGENICITY TO HUMANS. Herbologia, 12(2), 55-60.

Mathiassen, S. K., Bak, T., Christensen, S., & Kudsk, P. (2006). The Effect of Laser Treatment as a Weed Control Method. Biosystems Engineering, 95(4), 497-505. doi:http://dx.doi.org/10.1016/j.biosystemseng.2006.08.010

Melander, B., & Kristensen, J. K. (2011). Soil steaming effects on weed seedling emergence under the influence of soil type, soil moisture, soil structure and heat duration. Annals of Applied Biology, 158(2), 194-203. doi:10.1111/j.1744-7348.2010.00453.x

Nishimura, A., Asai, M., Shibuya, T., Kurokawa, S., & Nakamura, H. (2015). A steaming method for killing weed seeds produced in the current year under untilled conditions. Crop Protection, 71, 125-131. doi:http://dx.doi.org/10.1016/j.cropro.2015.02.015

Owen, M., Walsh, M., Llewellyn, R., & Powles, S. (2007). Widespread occurrence of multiple herbicide resistance in Western Australian annual ryegrass (Lolium rigidum) populations. Australian Journal of Agricultural Research, 58(7), 711-718.

Parish, S. (1990). A review of non-chemical weed control techniques. Biological Agriculture and Horticulture, 7(2), 177-1137. doi:10.1080/01448765.1990.9754540

Peighambarzadeh, S. Z., Safi, S., Shahtaheri, S. J., Javanbakht, M., & Forushani, A. R. (2011). Presence of Atrazine in the Biological Samples of Cattle and Its Consequence Adversity in Human Health. Iranian Journal of Public Health, 40(4), 112-121.

Price, A., & Kelton, J. (2011). Weed Control in Conservation Agriculture In M. Larramendy (Ed.), Herbicides, Theory and Applications (pp. 3-16). Rijeka, Croatia: InTech.

Raffaelli, M., Martelloni, L., Frasconi, C., Fontanelli, M., Carlesi, S., & Peruzzi, A. (2016). A prototype band-steaming machine: Design and field application. Biosystems Engineering, 144, 61-71. doi:http://dx.doi.org/10.1016/j.biosystemseng.2016.02.001

Rask, A. M., Larsen, S. U., Anderson, C., & Kristofferson, P. (2013). Determining treatment frequency for controlling weeds on traffic islands using chemical and non-chemical weed control. Weed Research, 53, 249-258.

Saunders, C., Godwin, R. J., & O'Dogherty, M. J. (2000). Prediction of Soil Forces Acting on Mouldboard Ploughs. Paper presented at the 4th International Conference on Soil Dynamics, Adelaide, Australia.

Storeheier, K. (1994). Basic investigations into flaming for weed control. Acta Horticulturae, 372, 195-204.

Tabatabaeefar, A., Emamzadeh, H., Varnamkhasti, M. G., Rahimizadeh, R., & Karimi, M. (2009). Comparison of energy of tillage systems in wheat production. Energy, 34(1), 41-45. doi:https://doi.org/10.1016/j.energy.2008.09.023

Troudi, A., Sefi, M., Ben Amara, I., Soudani, N., Hakim, A., Zeghal, K. M., et al. (2012). Oxidative damage in bone and erythrocytes of suckling rats exposed to 2,4-dichlorophenoxyacetic acid. Pesticide Biochemistry and Physiology, 104(1), 19-27. doi:http://dx.doi.org/10.1016/j.pestbp.2012.06.005

Vollmer, M. (2004). Physics of the microwave oven. Physics Education, 39(1), 74-81.

Wickerham, E. L., Lozoff, B., Shao, J., Kaciroti, N., Xia, Y., & Meeker, J. D. (2012). Reduced birth weight in relation to pesticide mixtures detected in cord blood of full-term infants. Environment International, 47(0), 80-85. doi:http://dx.doi.org/10.1016/j.envint.2012.06.007

Zoz, F. M., & Grisso, R. D. (2003). Traction and Tractor Performance. Paper presented at the Agricultural Equipment Technology Conference, Louisville, Kentucky USA.

## 4.7 Appendix A

Determining the peak radiation wavelength for any temperature:

$$\frac{d\rho}{d\lambda} = \frac{2c^3h^2e^{\frac{hc}{\lambda kT}}}{kT\lambda^7\left(e^{\frac{hc}{\lambda kT}} - 1\right)^2} - \frac{10c^2h}{\lambda^6\left(e^{\frac{hc}{\lambda kT}} - 1\right)} = 0$$

$$\frac{2c^3h^2e^{\frac{hc}{\lambda kT}}}{kT\lambda^7\left(e^{\frac{hc}{\lambda kT}} - 1\right)^2} = \frac{10c^2h}{\lambda^6\left(e^{\frac{hc}{\lambda kT}} - 1\right)}$$

Rearranging gives:

$$\lambda = \frac{2h^2c^3e^{\frac{hc}{\lambda kT}}\left(e^{\frac{hc}{\lambda kT}} - 1\right)}{10c^2hkT\left(e^{\frac{hc}{\lambda kT}} - 1\right)^2}$$

Or

$$\lambda = \frac{hc \cdot e^{\frac{hc}{\lambda kT}}}{5kT\left(e^{\frac{hc}{\lambda kT}} - 1\right)}$$

At normal temperatures $e^{\frac{hc}{\lambda kT}} \sim 10^{20}$ therefore:

$$\lambda \approx \frac{hc}{5kT}$$

## 4.8 Appendix B

Determining the total radiated power from a black body:

$$\mathrm{P} = 2\mathrm{hc}^2 \int_0^{\infty} \frac{1}{\lambda^5\left\{e^{\frac{\mathrm{hc}}{\lambda \mathrm{kT}}} - 1\right\}} \cdot d\lambda$$

Let $u = \frac{\mathrm{hc}}{\lambda \mathrm{kT}}$; $\lambda = \frac{\mathrm{hc}}{\mathrm{ukT}}$ and $d\lambda = \frac{-\mathrm{hc}}{\mathrm{u}^2\mathrm{kT}} \cdot du$

Substituting into the previous equation gives:

$$P = 2hc^2 \int_0^{\infty} \frac{\frac{-hc}{u^2 kT}}{\left(\frac{hc}{ukT}\right)^5 \{e^u - 1\}} \cdot du$$

Rearranging gives:

$$P = 2hc^2 \left(\frac{kT}{hc}\right)^4 \int_0^{\infty} \frac{u^3}{\{e^u - 1\}} \cdot du$$

Evaluating the integral gives:

$$P = 2hc^2 \left(\frac{kT}{hc}\right)^4 \frac{\pi^4}{15}$$

Rearranging gives:

$$P = \frac{2\pi^5 k^4}{15h^3c^2} T^4$$

This can be simplified to:

$$P = \sigma T^4$$

Where σ is the referred to as the Stefan-Boltzmann constant

In the case of a normal object the power transfer is reduced by a factor ε, depending on the properties of the object's surface; therefore, the power transfer is:

$$P = \varepsilon \sigma T^4$$

# 5 A Brief Review of Microwave Heating

## 5.1 Introduction

Microwave frequencies occupy the portion of the electromagnetic spectrum (300MHz to 300GHz) that lies between VHF radio-waves and thermal infrared. Their application falls into two categories, depending on whether the wave is used to transmit information or energy. The first category includes terrestrial and satellite communication links, radar, radio-astronomy, microwave thermography, material permittivity measurements, and so on (Adamski & Kitlinski, 2001). The second category of applications is associated with microwave heating and wireless power transmission. In the case of microwave heating, there is usually no signal modulation and the electromagnetic wave interacts directly with solid or liquid materials.

"*It has long been known that an insulating material can be heated by applying energy to it in the form of high frequency electromagnetic waves*" (Metaxas & Meredith, 1983, pp. 5). Industrial microwave heating has been used since the 1940's (Metaxas & Meredith, 1983, pp. 5). The initial experiments with microwave heating were conducted by Dr. Percy Spencer in 1946, following a serendipitous accident while he was testing a magnetron (Gallawa, 1998). Although Spencer was not the first to observe that microwave energy could impart heat to materials, he was the first to systematically study it. Since then many heating, drying, thawing (Liu, Marchant, Turner, & Vegh, 2003) and medical applications (Bond, Li, Hagness, & Van Veen, 2003) have been developed.

One key benefit of microwave heating, over conventional convective heating, is speed. The origin of this speed is the volumetric interactions between the microwave's electric field and the material. In contrast, convective heat transfer propagates from the surface into the material, with the final temperature profile depending on the material's thermal diffusion properties (Holman, 1997) and the influence of moisture transport, which often hinders the convective heating process (Crank, 1979).

The factors that contribute to microwave heating include: the physical and chemical structure of the heated material; the frequency of the microwaves (Van Remmen, Ponne, Nijhuis, Bartels, & Herkhof, 1996); in some cases, such as wood, the orientation of the electrical field relative to the structure of the dielectric material (Torgovnikov, 1993, pp. 13-17); reflections from the inter-facial surface of the heated material (Adamski & Kitlinski, 2001); electric field strength (Van Remmen et al., 1996); the geometry of the microwave applicator (Metaxas & Meredith, 1983); the geometry, size, electrical and thermal properties of the dielectric material (Brodie, 2008; Perre & Turner, 1999; Zhao, Turner, & Torgovnikov, 1998); the exposure time; and the moisture content of the dielectric material (Crank, 1979; Torgovnikov, 1993). This chapter briefly explores electromagnetic heating with a focus on some of the

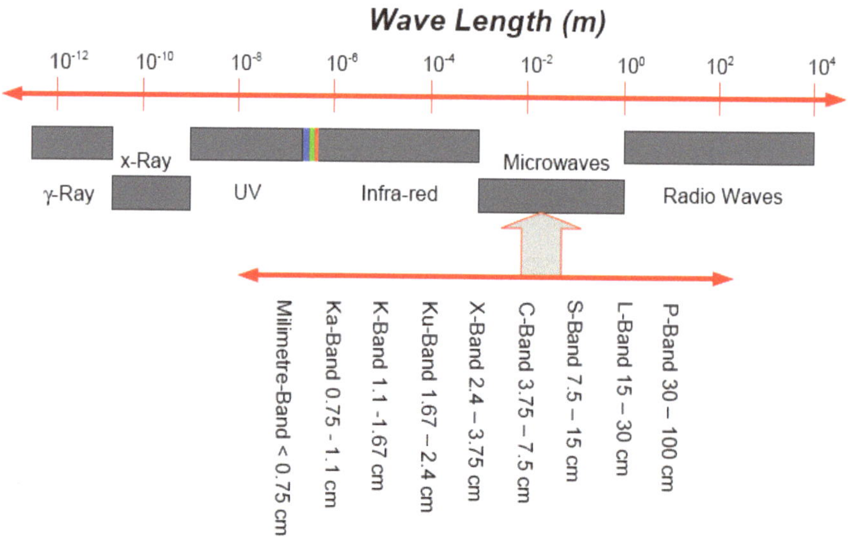


**Figure 5.1:** The electromagnetic spectrum, showing the microwave range in further detail.

devices (applicators) used to impose microwave fields onto dielectric materials and some applications of electromagnetic heating in biological systems.

## 5.2 Microwave Frequency and its Influence over Microwave Heating

Microwave frequencies occupy portions of the electromagnetic spectrum between 300 MHz to 300 GHz. The full range of microwave frequencies is further subdivided into various bands, as indicated in Figure 5.1.

Because microwaves are also used in the communication, navigation and defence industries, their use in thermal heating is restricted to a small subset of the available frequency bands. Commonly used frequencies include 434 ± 1 MHz, 922 ± 4 MHz, 2450 ± 50 MHz and 5800 ± 75 MHz (Commonwealth Department of Transport and Communications, 1991; International Telecommunication Union, 2004). In Australia, these frequencies have been set aside for Industrial, Scientific and Medical (ISM) applications (Commonwealth Department of Transport and Communications, 1991; International Telecommunication Union, 2004). All these frequencies interact to some degree with moist materials.

Microwave heating depends on the ability of the microwave's electric field to polarise dipolar molecules (Metaxas & Meredith, 1983). A dipole is essentially two equal and opposite charges separated by a finite distance. An example of this is the covalent bonds in a water molecule, giving the water molecule a dipole movement. Water is the typical case of a non-symmetric molecule, because there is a tendency for the larger oxygen atom to hold the two valance electrons more often than the two hydrogen atoms; therefore, inducing a slight negative change to the oxygen atom and a slight positive charge to the two hydrogen atoms. When this is coupled with the angular structure of the water molecule, there is a net positive end to the molecule and a net negative end. This results in a classic dipolar structure. Dipoles may be a natural feature of the dielectric or they may be induced (Kelly & Rowson, 1995). Distortion of the electron cloud around non-polar molecules or atoms through the presence of an external electric field can also induce a temporary dipole movement.

The interaction with an oscillating external electric field associated with the microwave energy induces torque on polar molecules and the resulting movement generates friction inside the dielectric, which is dissipated as heat. Depending on the frequency, the dipole may move in time with the electric field, lag behind it, or remain apparently unaffected (Chaplin, 2004). When the molecular dipole movement lags the applied field, interactions between the dipole and the field lead to energy dissipation within the material and heating. The extent of heating depends on the phase difference between the applied fields and the dipole moment of the molecules. The ease with which dipole movement occurs depends on the viscosity and the mobility of the electron clouds within the molecule (Chaplin, 2004). In the case of water these, in turn, depend on the strength and extent of the hydrogen bonded networks within the liquid phase (Chaplin, 2004). In free liquid water this movement occurs at GHz frequencies whereas in more restricted 'bound' water it occurs at MHz frequencies and in ice it occurs at kHz frequencies (Chaplin, 2004).

## 5.3 Electric Field Interactions with the Material and Their Influence over Microwave Heating

Coupled with these polarisation effects, a dielectric can also exhibit direct conduction within the material, as charges are displaced by the applied fields (Metaxas & Meredith, 1983). The combination of polarisation and conduction gives rise to displacement currents within the material (Torgovnikov, 1993). The complex displacement current can be resolved into a reactive component and a real component. To determine the current density in a dielectric material using Maxwell's equations and the physics of dielectric materials require the introduction of a complex dielectric constant $\varepsilon^*$ (Debye, 1929).

Debye (1929) deduced the well-known equation for the complex dielectric constant as:

$$\varepsilon^* = \varepsilon' + j\varepsilon'' = \varepsilon_\infty + \frac{\varepsilon_s - \varepsilon_\infty}{1 + j\omega\tau_r} \tag{5.1}$$

The relative dielectric constant ε' expresses the material's ability to store electrical energy (Singh and Heldman, 1993 pp. 208) and thus represents the reactive nature of the material's electrical properties (Giancoli, 1989; Smith, 1976). In particular, ε' influences the wave impudence of the space occupied by the dielectric causing reflections at the inter-facial boundary between the air and the dielectric material. Changes in wave impedance also cause refraction of the wave due to the change in the propagation velocity of the microwave within the dielectric material compared with its velocity in air or vacuum (Montoro, Manrique, & Gonzalez-Reviriego, 1999).

The dielectric loss ε" represents the resistive nature of the material's electrical properties (Giancoli, 1989; Smith, 1976). Resistive losses within the medium reduce the amplitude of the microwave field and generate heat inside the material.

The dielectric properties of most materials are temperature, frequency and moisture dependent. For example, Torgovnikov (1993) states that macromolecules such as cellulose, hemi-cellulose and lignin, which make up the wood cell wall, are also subject to dipole polarisation. This is associated with the displacement of polar groups such as OH and $CH_2OH$ relative to the motionless parts of these macromolecules. In spite of this, dry wood, which is basically a mixture of these macromolecules, air, and bound water, does not interact very strongly with microwaves (Chaplin, 2004; Torgovnikov, 1993). On the other hand, free liquid water in wood structures such as tracheids and vessels interacts strongly with microwaves (Chaplin, 2004) and has been described as the "*key substance attenuating microwaves*" during microwave heating in many natural materials (Zielonka & Gierlik, 1999).

Figure 5.2 shows the frequency and temperature dependency of the dielectric properties of free liquid water. It is interesting to note that the maximum dielectric loss occurs at much higher frequencies than those which are normally reserved for industrial microwave applications; however, the loss factor significantly increases with increasing concentrations of dissolved solids in the water, particularly salts. This is shown in Figure 5.3.

Because water plays such an important role in many organic systems, the dielectric properties of these materials is dependent on the water content of the samples (Figure 5.4).

In the particular case of anisotropic materials, such as wood, the orientation of the electric field vector has a significant effect on the dielectric properties of the material (Torgovnikov, 1993, pp. 13-17). Wood grain has a cylindrical geometry; therefore, within the wood structure there are three primary coordinates, corresponding to the normal cylindrical coordinate system. These are the radial direction running from the

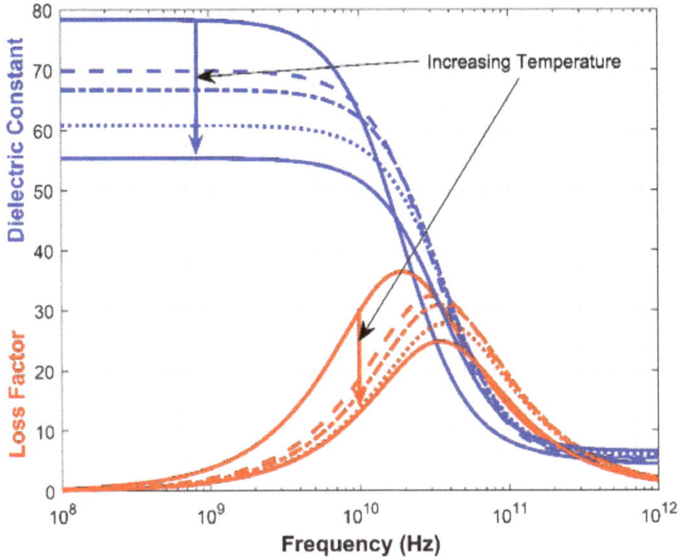


**Figure 5.2:** Dielectric constant and dielectric loss of water between 0°C and 100°C, the arrows showing the effect of increasing temperature.

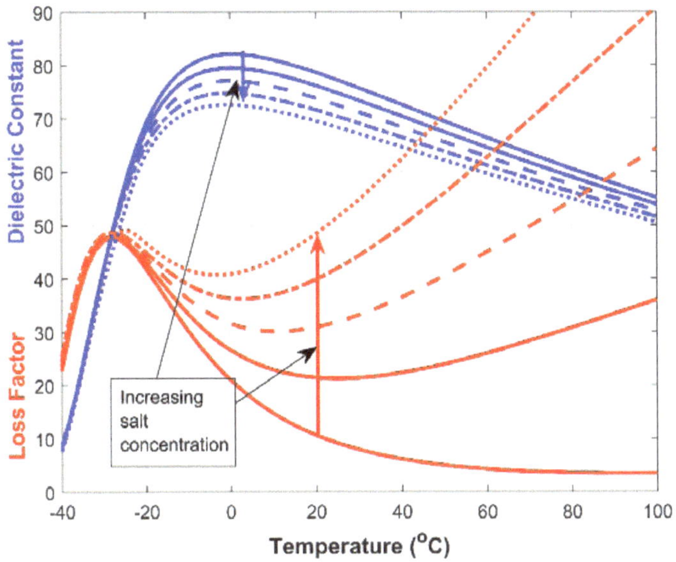


**Figure 5.3:** Dielectric constant and Loss Factor at 2.45 GHz for different parts per thousand w/w (ppt) salinity for the range for -20°C ~ +40°C.

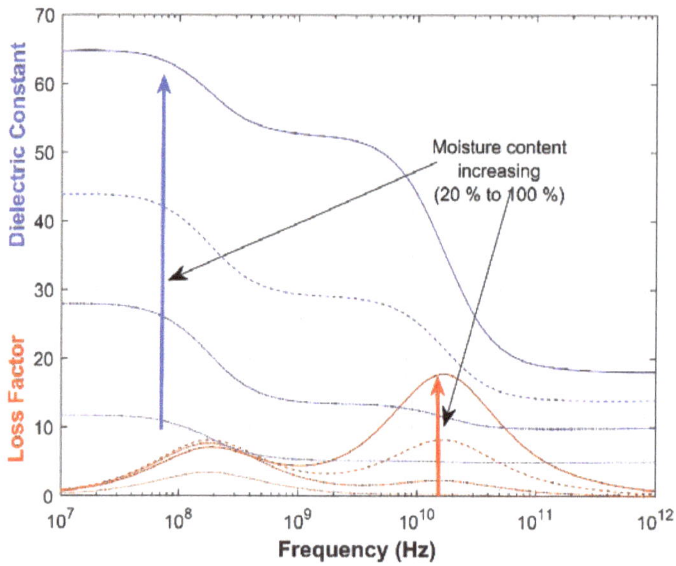


**Figure 5.4:** Dielectric properties of vegetative materials as a function of frequency and moisture content.

pith to the cambium, the tangential direction which is tangent to the growth rings and the longitudinal direction running along the length of the grain.

The power transmitted by an electromagnetic wave is proportional to the square of the electric field's magnitude (Giancoli, 1989). The electric field strength within a dielectric medium depends on the electrical field strength at the inter-facial surface between the material and the air, the reflection coefficient of this inter-facial surface, the geometry of the microwave applicator, the geometry of the material itself and the internal attenuation of the electric field with distance from the surface.

## 5.4 Challenges Associated with Microwave Research

According to McNamee and Chauhan (2009), there are several challenges when attempting to conduct high-quality RF and microwave research. Of primary importance is the interaction between the radiation and matter. If the effects of electromagnetic fields are of interest, the rate at which heat energy is applied to the sample must not exceed the rate at which it is removed; otherwise the temperature within the sample will rise and thermal confounding of the study may occur. If electromagnetic energy is applied to biological samples at a low rate (e.g. < 1 W $kg^{-1}$) or for a short duration, then passive cooling may permit the tissue/body/sample temperature to remain within a

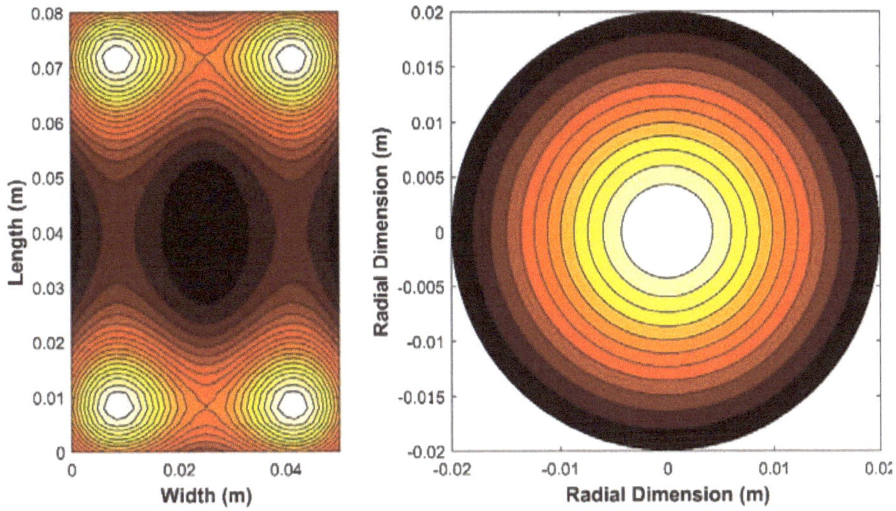


**Figure 5.5:** Temperature distribution in (left) cross section of a rectangular block and (right) cross section of a sphere with dielectric properties of e' = 9.6 and e" = 2.5 after 120 seconds of microwave heating at 2.45 GHz, based on models derived by Brodie (2008).

normal physiological range (McNamee & Chauhan, 2009). However, if higher specific absorption rates (>1 W $kg^{-1}$) are employed, then some form of active cooling mechanism may be required to ensure that excessive sample heating does not occur.

Another significant challenge is the non-homogeneity of energy absorption within the sample (McNamee & Chauhan, 2009). The development of hot-spots due to the nature of microwave heating has been an ongoing concern for many researchers (Metaxas & Meredith, 1983). The temperature distribution in a material undergoing RF or microwave heating is dependent on several factors; however, the geometry of the heated material itself tends to focus the electromagnetic energy into certain locations within the body of the heated object (Brodie, 2008; McNamee & Chauhan, 2009). Figure 5.5 illustrates the focusing effect of microwave heating in rectangular blocks and spheres.

A related problem is the manifestation of thermal runaway in these hot-spots. Thermal runaway, which manifests itself as a sudden temperature rise due to small increases in the applied microwave power, is very widely documented (Nelson, Wake, Chen, & Balakrishnan, 2001; Vriezinga, 1998; Zielonka & Dolowy, 1998). It has also been reported after some time of steady heating at fixed power levels and is usually attributed to temperature dependent dielectric and thermal properties of the material.

In addressing the phenomenon of thermal runaway, Vriezinga (1998) used analytical solutions to the differential equations that describe heat diffusion

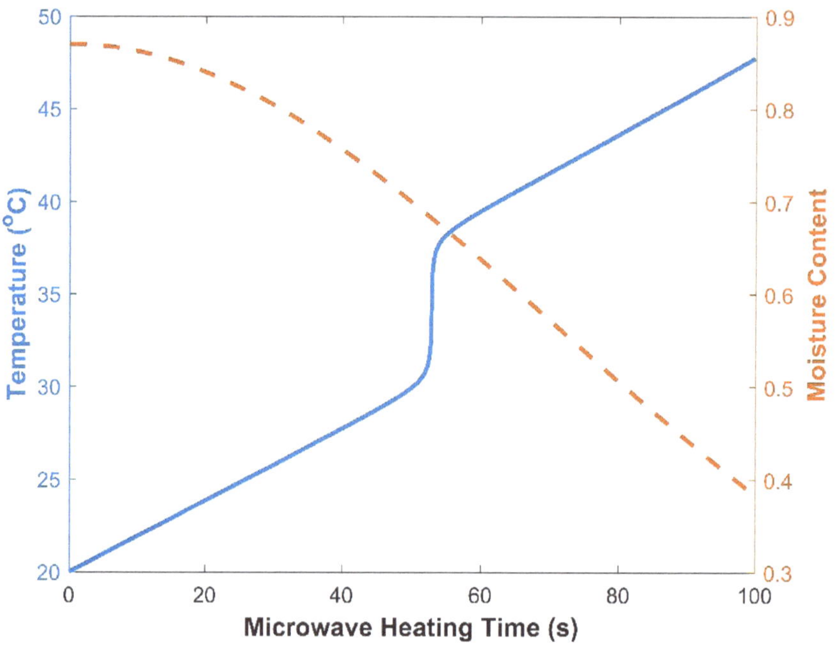


**Figure 5.6:** Example of thermal runaway in plant material dur to changing dielectric properties as the sample dries out.

(independent of moisture movement) in isothermal media to obtain S-shaped temperature versus microwave power curves. Hill and Marchant (1996) also developed S-shaped temperature versus microwave power curves during their investigation of microwave heating. They describe these curves as a multi-valued function of microwave power in which the upper and lower arms are stable, but the central arm is unstable. As the power increases from zero, the temperature stays on the lower arm until a critical power level is reached; then an infinitesimal increase in power will cause the temperature to jump to the upper arm. If the power is decreased the temperature will remain on the upper arm until a second critical power value is reached; then the temperature abruptly falls to the lower arm (Hill & Marchant, 1996). Liu et al. (2003) reiterate this interpretation of these curves attributing the sudden jump in temperature to thermal runaway.

Figure 5.6 shows the temperature and moisture content in the centre of a plant stem of 10 mm diameter heated with microwave energy. In this case, the application of microwave heating dries the sample, which ultimately reduces the dielectric properties of the sample. As the dielectric properties reach a threshold value, the field can resonate inside the material and cause a sudden jump in temperature. Vriezinga

(1998) also suggested that thermal runaway is a result of field resonance in the heated object.

## 5.5 Non-Thermal Effects

Various authors have proposed that changes in thermodynamic parameters under microwave irradiation, compared with those predicted by conventional heating, are caused by the "microwave effect" (De la Hoz, Díaz-Ortiz, & Moreno, 2007). Microbiological studies involving microwave irradiation have resulted in the following two conflicting conclusions; cell death was solely the result of heat produced by microwave irradiation; death was due to not only heat but also microwave electric field intensity (Banik, Bandyopadhyay, & Ganguly, 2003). The existence or otherwise of a "microwave effect" or non-thermal effect of microwave treatment is controversial.

Microwave photons have energies of the order of $10^{-5}$ eV (Vollmer, 2004). Simple estimates easily show that the number of microwave photons within a commercial oven is orders of magnitude too small to establish multiphoton dissociation or ionization of the processed materials (Vollmer, 2004); therefore the probability that microwave processing is having a non-thermal effect on the thermodynamics of the system is small.

One possible explanation for the observed changes in thermodynamic parameters, which have been observed in experimental work, is heat and vapour coupling (Brodie, 2007). Very rapid heating and drying during microwave processing of moist materials have been widely reported (Ni, Datta, & Parmeswar, 1999; Torgovnikov & Vinden, 2009; Vinden & Torgovnikov, 2000; Zielonka & Dolowy, 1998). The movement of hot moisture through the material, under the influence of microwave heating, effectively increases the thermal diffusivity and drying rate of the system. It is reasonable to expect that the thermal diffusivity and drying rates are linked to the applied microwave power.

In experiments using wool fibres, described by Cassie, King and Baxter (1940 in Crank, 1979), the isothermal moisture diffusion coefficient predicted that moisture equilibrium should be reached within seconds of a sudden change in external humidity; however, their experiments demonstrated that equilibrium was only reached after an hour or more of exposure to the new external conditions (Crank, 1979). Henry (1948) explored this phenomenon and deduced that there was strong coupling between the thermal behaviour and moisture movement in these porous textiles. His work revealed that the combined processes of heat and vapour diffusion are equivalent to the independent diffusion of two quantities, each of which is a linear function of vapour concentration and temperature. The diffusion coefficients of these two quantities are always such that one is much less and the other much greater than would be observed, were vapour and heat diffusion not coupled together; therefore,

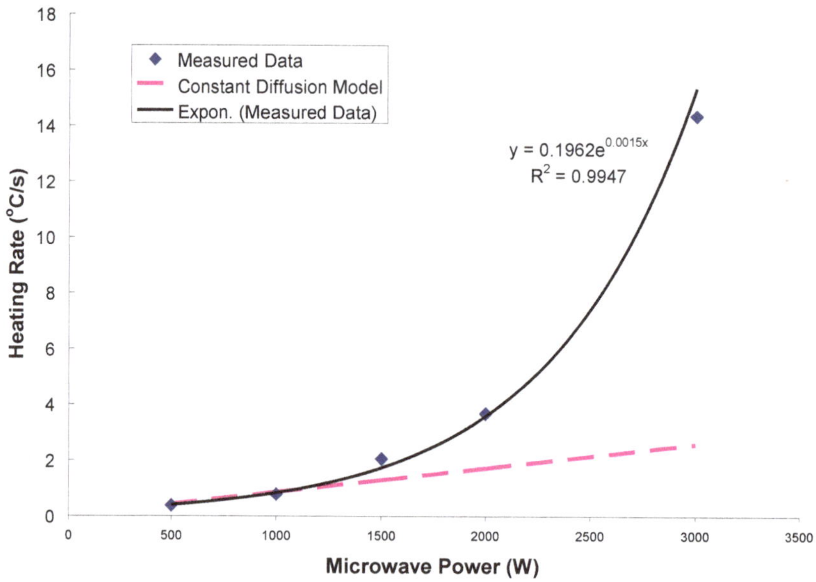


**Figure 5.7:** Heating rate in 25 mm by 25 mm Eucalyptus regnans samples, heated in a wave-guide, as a function of applied power (Dotted line represents the response that constant diffusion should produce).

the independent solution of the heat and vapour diffusion equations is inadequate to describe their combined influence (Henry, 1948).

The diffusion coefficient for the slower quantity of the coupled system is always less than either the isothermal moisture diffusion constant or the constant vapour concentration coefficient for heat diffusion, whichever is less, but never by more than one half (Henry, 1948). The faster diffusion coefficient may be many times greater than either of the independent diffusion constants (Henry, 1948).

Henry (1948) presents a nomogram relating the fast diffusion coefficient to the default diffusion coefficient for the material at 20°C and 65% relative humidity. This nomogram can be used to forecast the thermal diffusivity of the system under different conditions. For example, if the relative humidity remains constant and the temperature of some part of the system rapidly rises to 55°C, Henry's nomogram suggests that the diffusion coefficient for the faster wave will be about 7.5 times higher than the standard thermal diffusivity of the material. Figure 5.7 shows the relationship between heating rates and applied microwave power during microwave heating experiments involving *Eucalyptus regnans* wood samples, which were conducted at the University of Melbourne. It would be tempting to attribute the non-linear response of the samples to a "microwave effect"; however, it may be better to regard this as a progressively stronger coupling between the thermal and vapour diffusion in the samples, as the applied microwave power increases.

## 5.6 References

Adamski, W., & Kitlinski, M. (2001). On measurements applied in scientific researches of microwave heating processes. Measurement Science Review, 1(1), 199-203.

Banik, S., Bandyopadhyay, S., & Ganguly, S. (2003). Bioeffects of microwave--a brief review. Bioresource Technology, 87(2), 155-159.

Bond, E. J., Li, X., Hagness, S. C., & Van Veen, B. D. (2003). Microwave imaging via space-time beamforming for early detection of breast cancer. IEEE Transaction on Antennas and Propagation, 51(8), 1690-1705.

Brodie, G. (2007). Simultaneous heat and moisture diffusion during microwave heating of moist wood. Applied Engineering in Agriculture, 23(2), 179-187.

Brodie, G. (2008). The influence of load geometry on temperature distribution during microwave heating. Transactions of the American Society of Agricultural and Biological Engineers, 51(4), 1401-1413.

Chaplin, M. (2004). Water Structure and Behavior. Retrieved from http://www.lsbu.ac.uk/water/index.html

Commonwealth Department of Transport and Communications. (1991). Australian Radio Frequency Spectrum Allocations. Commonwealth Department of Transport and Communications.

Crank, J. (1979). The Mathematics of Diffusion. Bristol: J. W. Arrowsmith Ltd.

De la Hoz, A., Díaz-Ortiz, A., & Moreno, A. (2007). REVIEW ON NON-THERMAL EFFECTS OF MICROWAVE IRRADIATION IN ORGANIC SYNTHESIS. Journal of Microwave Power & Electromagnetic Energy, 41(1), 1.44.

Debye, P. (1929). Polar Molecules. New York: Chemical Catalog.

Gallawa, J. C. (1998). The Complete Microwave Oven Service Handbook: Operation, Maintenance, Troubleshooting and Repair. Englewood Cliffs: Prentice Hall.

Giancoli, D. C. (1989). Physics for Scientists and Engineers (2nd ed.). New Jersey: Prentice Hall.

Henry, P. S. H. (1948). The diffusion of moisture and heat through textiles. Discussions of the Faraday Society, 3, 243-257.

Hill, J. M., & Marchant, T. R. (1996). Modelling Microwave Heating. Applied Mathematical Modeling, 20(1), 3-15.

Holman, J. P. (1997). Heat Transfer (8th ed.). New York: McGraw-Hill.

International Telecommunication Union. (2004). Spectrum Management for a Converging World: Case Study on Australia. Retrieved from

Kelly, R. M., & Rowson, N. A. (1995). Microwave reduction of oxidised ilmenite concentrates. Minerals Engineering, 8(11), 1427–1438.

Liu, B., Marchant, T. R., Turner, I. W., & Vegh, V. (2003). A comparison of semi-analytical and numerical solutions for the microwave heating of lossy material in a three-dimensional waveguide. Paper presented at the Third World Congress on Microwave and Radio Frequency Applications, Sydney.

McNamee, J. P., & Chauhan, V. (2009). Radiofrequency radiation and gene/protein expression: A review. Radiation Research, 172(3), 265-287.

Metaxas, A. C., & Meredith, R. J. (1983). Industrial Microwave Heating. London: Peter Peregrinus.

Montoro, T., Manrique, E., & Gonzalez-Reviriego, A. (1999). Measurement of the refracting index of wood for microwave radiation. Holz als Roh- und Werkstoff, 57(4), 295 -299.

Nelson, M. I., Wake, G. C., Chen, X. D., & Balakrishnan, E. (2001). The multiplicity of steady-state solutions arising from microwave heating. I. Infinite Biot number and small penetration depth. The ANZIAM Journal, 43(1), 87-103.

Ni, H., Datta, A. K., & Parmeswar, R. (1999). Moisture loss as related to heating uniformity in microwave processing of solid foods. Journal of Food Process Engineering, 22(5), 367-382.

Perre, P., & Turner, I. W. (1999). The use of numerical simulation as a cognitive tool for studying the microwave drying of softwood in an over-sized wave-guide. Wood Science and Technology, 33(6), 445-464.

Smith, R. J. (1976). Circuits, Divices and Systems (3rd ed.). New York: Wiley International.

Torgovnikov, G., & Vinden, P. (2009). High-intensity microwave wood modification for increasing permeability. Forest Products Journal, 59(4), 84-92.

Torgovnikov, G. I. (1993). Dielectric Properties of Wood and Wood-Based Materials. Berlin: Springer-Verlag.

Van Remmen, H. H. J., Ponne, C. T., Nijhuis, H. H., Bartels, P. V., & Herkhof, P. J. A. M. (1996). Microwave Heating Distribution in Slabs, Spheres and Cylinders with Relation to Food Processing. Journal of Food Science, 61(6), 1105-1113.

Vinden, P., & Torgovnikov, G. (2000). The physical manipulation of wood properties using microwave. Paper presented at the International Conference of IUFRO, Tasmania, Australia.

Vollmer, M. (2004). Physics of the microwave oven. Physics Education, 39(1), 74-81.

Vriezinga, C. A. (1998). Thermal runaway in microwave heated isothermal slabs, cylinders, and spheres. Journal of Applied Physics, 83(1), 438 -442.

Zhao, H., Turner, I. W., & Torgovnikov, G. (1998). An experimental and numerical investigation of the microwave heating of wood. The Journal of Microwave Power and Electromagnetic Energy, 33(2), 121-133.

Zielonka, P., & Dolowy, K. (1998). Microwave Drying of Spruce: Moisture Content, Temperature and Heat Energy Distribution. Forest Products Journal, 48(6), 77-80.

Zielonka, P., & Gierlik, E. (1999). Temperature distribution during conventional and microwave wood heating. Holz als Roh- und Werkstoff, 57(4), 247-249.

# 6 A Brief History of Microwave Weed Control Research

## 6.1 Introduction

Interest in the effects of high frequency electromagnetic waves on biological materials dates back to the late 19th century (Ark & Parry, 1940), while interest in the effect of high frequency waves on plant material began in the 1920s (Ark & Parry, 1940). Many of the earlier experiments on plant material focused on the effect of radio frequencies (RFs) on seeds (Ark & Parry, 1940). In many cases, short exposure resulted in increased germination and vigour of the emerging seedlings (Nelson, Ballard, Stetson, & Buchwald, 1976; Nelson & Stetson, 1985; Tran, 1979); however, long exposure usually resulted in seed death (Ark & Parry, 1940; Bebawi et al., 2007; Brodie et al., 2009).

Davis et al. (1971; 1973) were among the first to study the lethal effect of microwave heating on seeds. They developed a set of prototypes, called "Zappers", which they tested in the field for their Company and federal and state researchers. Their final prototype, designated Zapper III, underwent tests underwent tests to provide the data necessary for the construction of the first semi-commercial prototype. In October 1971, the Company purchased all proprietary rights to a discovery made at Texas A&M University concerning the toxic effects of microwaves on plants Davis et al. (1971; 1973).

## 6.2 Pioneering Work

The discovery was the result of the efforts by Drs Merkle, Wayland, and Davis, who were originally professors in the Soil and Crop Sciences, Physics, and Range Science Departments, respectively, of Texas A&M University. The Company's first field prototype was named Zapper I. Zapper I was used in a cooperative testing program with US federal and state agricultural research agencies and with growers in Texas, California, Florida, New Mexico, Washington, Idaho, Nebraska, Arkansas, North Carolina, Georgia and Michigan. The Zapper I test program proved that microwaves could safely treat soil and be an effective herbicide. In addition, microwaves also proved to be toxic to nematodes, certain fungi, and to soil-borne insect pests. Further, the phenomenon of growth stimulation was first observed in plants which germinated in treated soil (Davis, 1974).

Following the initial Zapper I program, the Company built a second prototype, the Zapper III (Figure 6.1), which was used to determine the cost of Zapper treatments required to destroy various types of weed seeds under different soil conditions. The Zapper III program also experimented with different equipment configurations to determine the most efficient system design for commercial use (Davis, 1974). Both systems operated at a frequency of 2.45 GHz.

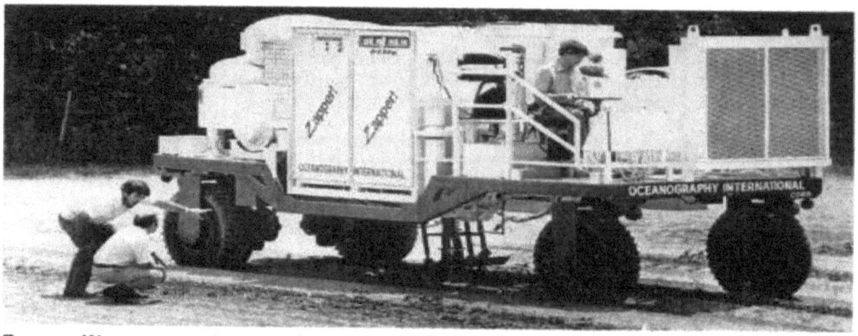


**Zapper III zaps weed and soil pests by discharging high-powered microwaves into the ground via a sled-shaped emitter. Developed by Oceanography International Corp., Zapper can increase crop yields without the risk of pesticide use.**

**Figure 6.1:** The Zapper III microwave prototype during field trials (Source: Anonymous, 1975).

A meta-study of published data (Menges & Wayland, 1974; Wayland, Merkle, Davis, Menges, & Robinson, 1975) reveals that microwave treatment of emerged weed plants, of eleven species, can be described by equations of the form (Figure 6.2):

$$S=a\bullet \operatorname{erfc}[b(\Psi-c)] \tag{6.1}$$

When the weed species are separated into categories of broad leafed and grasses, it appears that grasses require slightly more microwave energy to achieve treatment efficacy, compared with broad leafed plants (Figure 6.3).

It also became apparent that microwave treatment of the soil could inactivate weed seeds at various depths (Menges & Wayland, 1974; Wayland et al., 1975). The efficacy of the treatment depended on the soil type, the seed burial depth, the microwave treatment energy density and whether the soil had been irrigated prior to treatment (Figure 6.4). Irrigation prior to treatment resulted in shallower microwave heating; therefore, seed which were buried deeper in the soil profile were less affected by the microwave heating (Menges & Wayland, 1974; Wayland et al., 1975). The consensus from this data is that 300 – 500 J $cm^{-2}$ of microwave energy density at the soil surface, can control weeds and their seeds in the top 4 – 6 cm of soil. This is equivalent to between 30 and 50 GJ $ha^{-1}$ of microwave energy, making microwave treatment a little more energy expensive than steam treatment (see Chapter 4).

It is unclear, from the available literature, why this promising technology did not become more widely available as a commercial system. It is apparent that the ideas generated by this early work interest persisted into the 1990's, because Nelson (1996) used a theoretical argument to dismiss microwave soil treatment as a viable prospect for weed management. The high energy input required to achieve good weed and seed control was certainly a strong argument against the adoption of this technology.

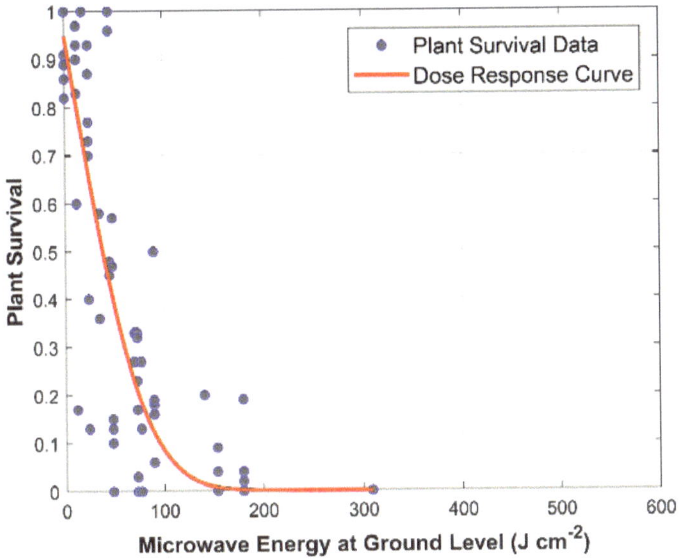


**Figure 6.2:** Response of 11 species of weed to microwave energy (Sources: Menges & Wayland, 1974; Wayland et al., 1975).

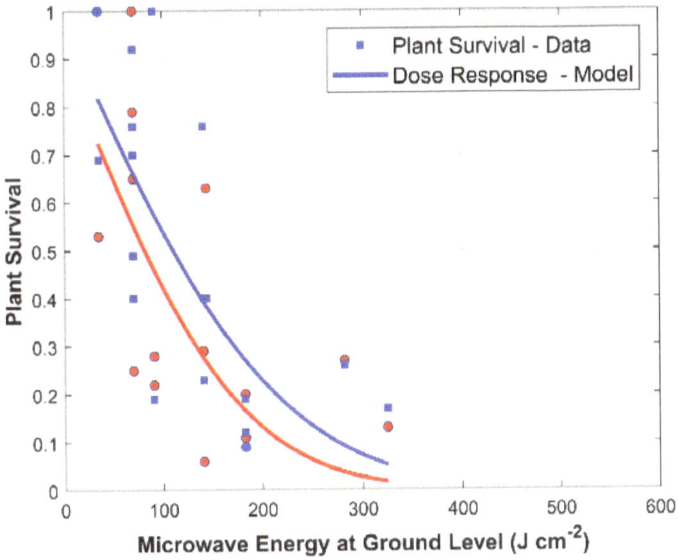


**Figure 6.3:** Response of grasses (blue) and broad-leafed weeds (red) to microwave energy (Sources: Menges & Wayland, 1974; Wayland et al., 1975).

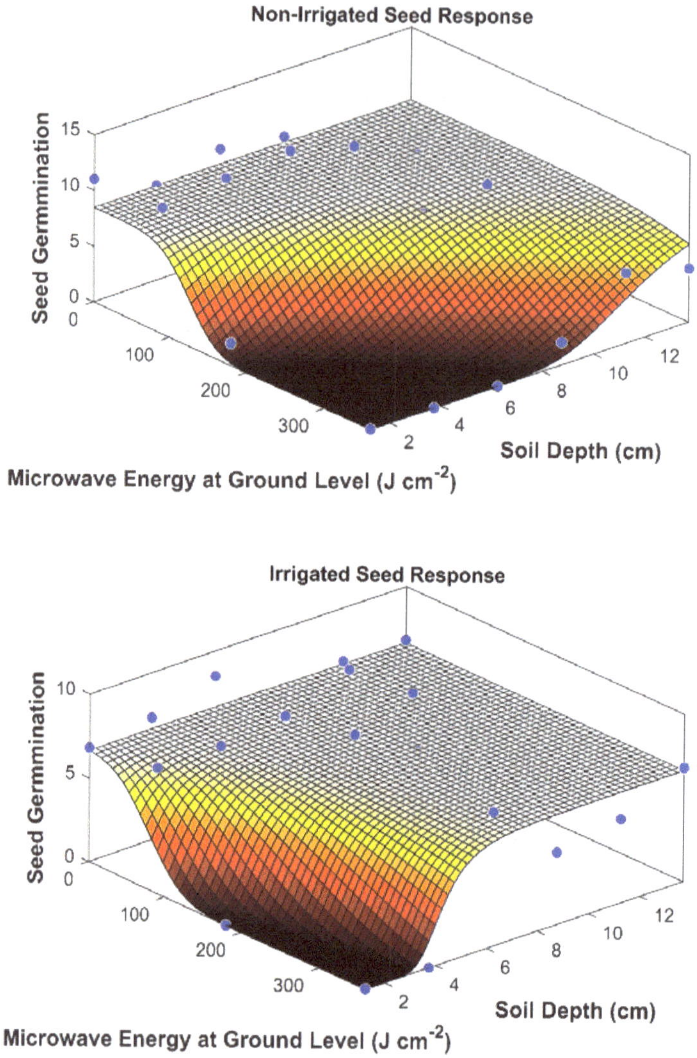


**Figure 6.4:** Response weed seeds in the soil to microwave energy, as a function of applied energy density, burial depth and irrigation status (Sources: Menges & Wayland, 1974; Wayland et al., 1975).

Despite this, then there has been ongoing research interest in microwave soil treatment and weed management. Table 6.1 lists a subset of the papers that have been published on these and related topics. The consensus from these studies is that: microwave treatment can kill plants; moderate microwave treatment can break dormancy in some hard-seeded species; and high energy microwave treatment can kill seeds in the soil.

**Table 6.1:** Literature addressing the application of microwave technology to seed and weed treatment.

| Paper Title | Reference |
|---|---|
| Douglas- fir tree seed germination enhancement using microwave energy | (Jolly & Tate, 1971) |
| Microwave processing of tree seeds | (Kashyap & Lewis, 1974) |
| Increasing legume seed-germination by VHF and microwave dielectric heating | (Nelson et al., 1976) |
| Effects of low-level microwave radiation on germination and growth rate in corn seeds | (Bigu-Del-Blanco, Bristow, & Romero-Sierra, 1977) |
| Effects of Microwave Energy on the Strophiole, Seed Coat and Germination of Acacia Seeds | (Tran, 1979) |
| The effect of microwave-energy on germination and dormancy of wild oat seeds | (Lal & Reed, 1980) |
| The Effect of Externally Applied Electrostatic Fields, Microwave Radiation and Electric Currents on Plants and Other Organisms, with Special Reference to Weed Control | (Diprose, Benson, & Willis, 1984) |
| Control of field weeds by microwave radiation | (Vela-Múzquiz, 1984) |
| Effect of microwave irradiation on germination and initial growth of mustard seeds | (Rao, Chakravarthy, & Panda, 1989) |
| Inhibition of weed seed germination by microwaves | (Barker & Craker, 1991) |
| A possibility of correction of vital processes in plant cell with microwave radiation | (Petrov, Moiseeva, & Morozova, 1991) |
| Microwave irradiation of seeds and selected fungal spores | (Cavalcante & Muchovej, 1993) |
| Response surface models to describe the effects and phytotoxic thresholds of microwave treatments on barley seed germination and vigour | (Stephenson, Kushalappa, Raghavan, & Mather, 1996) |
| Energy Efficient Soil Disinfestation by Microwaves | (Mavrogianopoulos, Frangoudakis, & Pandelakis, 2000) |
| Microwave effects on germination and growth of radish (*Raphanus sativus* L.) seedlings | (Scialabba & Tamburello, 2002) |
| Report on the Development of Microwave System for Sterilisation of Weed Seeds: Stage I – Feasibility | (Advanced Manufacturing Technologies, 2003) |
| Design, construction and preliminary tests of a microwave prototype for weed control | (Zanche, Amista, Baldoin, Beria, & Giubbolini, 2003) |
| Thermal effects of microwave energy in agricultural soil radiation | (Velazquez-Marti & Gracia-Lopez, 2004) |
| Influence of low-frequency and microwave electromagnetic fields on seeds | (Kalinin, Boshkova, Panchenko, & Kolomiichuk, 2005) |
| An improved microwave weed killer | (Vidmar, 2005) |

Continued Table 6.1: Literature addressing the application of microwave technology to seed and weed treatment.

| Paper Title | Reference |
|---|---|
| Observations on the potential of microwaves for weed control | (Sartorato, Zanin, Baldoin, & De Zanche, 2006) |
| Plant response to microwaves at 2.45 GHz. | (Skiles, 2006) |
| Germination Inhibition of Undesirable Seed in the Soil using Microwave Radiation | (Velazquez-Marti, Gracia-Lopez, & Marzal-Domenech, 2006) |
| Effect of microwave radiation on seed mortality of rubber vine (*Cryptostegia grandiflora* R.Br.), parthenium (*Parthenium hysterophorous* L.) and bellyache bush (*Jatropha gossypiifolia* L.) | (Bebawi et al., 2007) |
| Effects of microwave treatment on growth, photosynthetic pigments and some metabolites of wheat | (Hamada, 2007) |
| Microwave seed treatment reduces hardseededness in Stylosanthes seabrana and promotes redistribution of cellular water as studied by NMR relaxation measurements | (Anand, S, Joshi, Verma, & Kar, 2008) |
| Effect of microwave fields on the germination period and shoot growth rate of some seeds | (Monteiro, Mendiratta, & Capitão, 2008) |
| Germination of *Chenopodium Album* in Response to Microwave Plasma Treatment | (Sera, Stranak, Sery, Tichy, & Spatenka, 2008) |
| Work conditions for microwave applicators designed to eliminate undesired vegetation in a field | (Velazquez-Marti, Gracia-Lopez, & de la Puerta, 2008) |

## 6.3 Microwave Weed and Soil Treatment Patents

The long-standing interest in applying microwave technology to weed and soil treatment has resulted in many attempts to capture the intellectual property through various patents (Tab. 6.2). It is evident that some of these are the same invention; however, they have been patented in different parts of the world. Patents have included two main methods of soil treatment: in-situ treatment systems that do not disturb the soil (Clark & Kissell, 2003; Haller, 2002; Joines, 2009); and tunnel treatment systems which use some mechanical method to remove the top soil, pass it through a microwave treatment chamber or tunnel and then return the soil to its original position after treatment (Wall, 2009). The in-situ treatment systems use various antenna systems or multi-mode cavities (somewhat like half of a microwave oven that is open to the soil) to apply the microwave energy (For example: Clark & Kissell, 2003; Haller, 2002). Several of these patents claim to control other crop pests as well as weeds and their seeds in the soil (Grigorov, 2003; Haller, 2002; Joines, 2009). There are also several companies that have developed microwave based weed management technologies, but have chosen not to apply for a patent to protect their inventions. There will be others that the authors are not aware of. Some of these companies have developed mature technologies; however most have systems that are in the developmental stage.

## 6.4 Conclusion

It is clear from the number of papers, patents and other evidence that the basic principle of microwave weed management is of considerable interest and is well understood. Several system designs have been developed and protected; however, there is still scope to develop novel microwave applicator designs that better couple the microwave energy into the soil and weed plants. There is also opportunity to develop and implement better energy control systems that could reduce the energy required to achieve effective soil and weed treatment and automate the weed management process.

On the more cautionary side, in a theoretical argument based on the dielectric and physical properties of seeds and soils, Nelson (1996) demonstrated that using microwaves to selectively heat seeds in the soil "*cannot be expected.*" He stated that seed susceptibility to damage from microwave treatment is a purely thermal effect, resulting from soil heating and thermal conduction into the seeds. He concluded that microwave weed management was not viable; however, his arguments ignored any effects of herbicide resistance on crop yields.

**Table 6.2:** Patents which address or are associated with microwave weed and soil treatment.

| Publication Number | Priority Date | Filing Date | Date of Publication | Title |
|---|---|---|---|---|
| EP 0413847 A1<br>US4861956A<br>WO1991002548A1 | 17/10/1986 | 24/08/1989<br>17/10/1986<br>24/08/1989 | 27/02/1991<br>29/08/1989<br>7/03/1991 | **Microwave/steam sterilizer.** \| Mikrowellen-/Dampf-Sterilisator. \| Stérilisateur à micro-ondes et à vapeur. |
| US5287818A | 11/05/1993 | 11/05/1993 | 22/02/1994 | Method for killing soil pathogens with micro-wave energy |
| US5141059A | 27/02/1991 | 27/02/1991 | 25/08/1992 | Method and apparatus for controlling agricultural pests in soil |
| CA2299301 A1<br>DE69625089D1<br>DE69625089T2<br>EP0928134 A1<br>EP0928134 B1<br>US6237278B1<br>WO1998/007314 A1 | 16/08/1996<br><br><br><br><br><br>20/02/1995 | 15/08/1996<br>16/08/1996<br>16/08/1996<br>16/08/1996<br>16/08/1996<br>16/08/1996<br>16/08/1996 | 26/02/1998<br>9/01/2003<br>4/09/2003<br>14/07/1999<br>27/11/2002<br>29/05/2001<br>26/02/1998 | **Method and device for weed control** \| Procede et dispositif de desherbage \| Verfahren und vorrichtung zur unkrautbekämpfung |

Continued Table 6.2: Patents which address or are associated with microwave weed and soil treatment.

| Publication Number | Priority Date | Filing Date | Date of Publication | Title |
|---|---|---|---|---|
| DE 19850195 A1 | 22/10/1998 | 22/10/1999 | 4/05/2000 | **Method and device for killing wood-destroying animals** \| Procede et dispositif pour exterminer des parasites animaux dans le bois \| Verfahren und vorrichtung zum abtöten von tierischen schädlingen in holz |
| DE 59915075 D1 | | 22/10/1999 | 1/02/2007 | |
| EP 1158853 A1 | | 22/10/1999 | 5/12/2001 | |
| WO2000/024247 A1 | | 22/10/1999 | 4/05/2000 | |
| CA2372471A1 | 4/04/2000 | 3/04/2001 | 18/10/2001 | **Method and system for exterminating pests, weeds and pathogens** \| Procede et systeme d'extermination d'animaux nuisibles, de plantes nuisibles et d'agents pathogenes \| Verfahren und system zur vernichtung von ungeziefer, unkraut und pathogenen |
| CA2372471C | | 3/04/2001 | 11/12/2007 | |
| DE60114392D1 | | 3/04/2001 | 1/12/2005 | |
| DE60114392T2 | | 3/04/2001 | 27/07/2006 | |
| EP1272032A1 | | 3/04/2001 | 8/01/2003 | |
| EP1272032B1 | | 3/04/2001 | 26/10/2005 | |
| US20030037582A1 | | 3/04/2001 | 27/02/2003 | |
| US6647661B2 | | 3/04/2001 | 18/11/2003 | |
| WO2001/076362 A1 | | 3/04/2001 | 18/10/2001 | |
| US 6401637 B1 | 8/01/2001 | 15/06/2001 | 11/06/2002 | Microwave energy applicator |
| US 20020090268 | | 15/06/2001 | 11/07/2002 | |
| EP 1224863 A2 | 15/11/2001 | 15/11/2001 | 24/07/2002 | **Microwave disinfestation system for biological pests** \| Système de désinfection à micro-ondes pour lutte biologique \| Mikrowellendes Infektionsystem für biologische Schädlinsgbekämpfung |
| EP 1224863 A3 | | 15/11/2001 | 21/09/2005 | |
| US20040009092A1 | | 15/07/2002 | 15/01/2004 | |
| CA 2483749 A1 | 28/03/2002 | 27/03/2003 | 9/10/2003 | **Method and apparatus [device] for controlling pests found in the ground, in particular termites \|** Procede et dispositif pour lutter contre les animaux nuisibles vivant dans le sol, en particulier les termites \| Verfahren und Vorrichtung zur Bekämpfung von im Erdboden hausenden Schädlingen, insbesondere Termiten \| 防治在土壤中筑巢的有害动物特别是白蚁的方法和装置 |
| CN 1642414 A | | 27/03/2003 | 20/07/2005 | |
| DE 10213983 C1 | | 28/03/2002 | 13/11/2003 | |
| EP 1487263 A1 | | 27/03/2003 | 22/12/2004 | |
| US 20050039379 A1 | | 27/09/2004 | 24/02/2005 | |
| WO2003/081999 A1 | | 27/03/2003 | 9/10/2003 | |

Continued Table 6.2: Patents which address or are associated with microwave weed and soil treatment.

| Publication Number | Priority Date | Filing Date | Date of Publication | Title |
|---|---|---|---|---|
| US20030215354 A1 | 17/05/2002 | 16/09/2002 | 20/11/2003 | **Systems and methods for in situ soil sterilization, insect extermination and weed killing** \| Systemes et procedes de sterilisation des sols in situ, d'extermination des insectes et de desherbage |
| WO2003/099004 A2 | | 4/10/2002 | 4/12/2003 | |
| WO2003/099004 A3 | | 4/10/2002 | 17/06/2004 | |
| US20060186115A1 | 11/01/2005 | 11/01/2006 | 24/08/2006 | Microwave system and method for controlling the sterlization and infestation of crop soils |
| US20090232602A1 | | 22/09/2009 | 17/09/2009 | |
| US7601936B2 | | | 13/10/2009 | |
| US201220091123A1 | | | 19/04/2012 | |
| US20080149625A1 | 25/10/2006 | 25/10/2006 | 26/06/2008 | Device for soil sterilization, insect extermination, and weed killing using microwave energy |
| US7560673B2 | | 25/10/2006 | 14/07/2009 | |
| WO2008057215A2 | | 24/10/2007 | 15/05/2008 | **A device and method for soil sterilization, insect extermination, and weed killing using microwave energy** \| Dispositif et procédé de stérilisation de sol, d'extermination d'insectes et de désherbage par énergie micro-onde |
| WO2008057215A3 | | 24/10/2007 | 24/07/2008 | |
| US 20130212928A1 | 17/02/2012 | 13/02/2013 | 22/08/2013 | **Apparatus for using microwave energy for insect and pest control and methods thereof** \| Appareil d'utilisation de l'énergie micro-onde pour le contrôle des insectes et des animaux nuisibles et procédés associés |
| US 20150101239A1 | | 18/12/2014 | 16/04/2015 | |
| US 8943744 B2 | | 13/02/2013 | 3/02/2015 | |
| WO2013/123089 A1 | | 13/02/2013 | 22/08/2013 | |

## 6.5 References


Advanced Manufacturing Technologies. (2003). Report on the Development of Microwave System for Sterilisation of Weed Seeds: Stage I - Feasibility. Retrieved from Wollongong, NSW.

Anand, A., S, N., Joshi, D. K., Verma, A. P.-S., & Kar, A. (2008). Microwave seed treatment reduces hardseededness in Stylosanthes seabrana and promotes redistribution of cellular water as studied by NMR relaxation measurements. Seed Science and Technology, 37(1), 88-97.

Anonymous. (1975). Prize Winning Products from the IR-100 Show. Popular Science, January, 46.

Ark, P. A., & Parry, W. (1940). Application of High-Frequency Electrostatic Fields in Agriculture. The Quarterly Review of Biology, 15(2), 172-191.

Barker, A. V., & Craker, L. E. (1991). Inhibition of weed seed germination by microwaves. Agronomy Journal, 83(2), 302-305.

Bebawi, F. F., Cooper, A. P., Brodie, G. I., Madigan, B. A., Vitelli, J. S., Worsley, K. J., & Davis, K. M. (2007). Effect of microwave radiation on seed mortality of rubber vine (Cryptostegia grandiflora R.Br.), parthenium (Parthenium hysterophorous L.) and bellyache bush (Jatropha gossypiifolia L.). Plant Protection Quarterly, 22(4), 136-142.

Bigu-Del-Blanco, J., Bristow, J. M., & Romero-Sierra, C. (1977). Effects of low-level microwave radiation on germination and growth rate in corn seeds. Proceedings of the IEEE, 65(7), 1086-1088.

Brodie, G., Harris, G., Pasma, L., Travers, A., Leyson, D., Lancaster, C., & Woodworth, J. (2009). Microwave soil heating for controlling ryegrass seed germination. Transactions of the American Society of Agricultural and Biological Engineers, 52(1), 295-302.

Cavalcante, M. J. B., & Muchovej, J. J. (1993). Microwave irradiation of seeds and selected fungal spores. Seed Science and Technology, 21(1), 247-253.

Clark, W. J., & Kissell, C. W. (2003). United States Patent No. 20030215354A1.

Davis, F. (1974). New techniques in weed control via microwaves. Paper presented at the Western Session- Southeastern Nurserymen's Conferences Nacogdoches, Texas.

Davis, F. S., Wayland, J. R., & Merkle, M. G. (1971). Ultrahigh-Frequency Electromagnetic Fields for Weed Control: Phytotoxicity and Selectivity. Science, 173(3996), 535-537.

Davis, F. S., Wayland, J. R. and Merkle, M. G. (1973). Phytotoxicity of a UHF Electromagnetic Field. Nature, 241(5387), 291-292.

Diprose, M. F., Benson, F. A., & Willis, A. J. (1984). The Effect of Externally Applied Electrostatic Fields, Microwave Radiation and Electric Currents on Plants and Other Organisms, with Special Reference to Weed Control. Botanical Review, 50(2), 171-223.

Grigorov, G. R. (2003). United States Patent No. 20030037482A1.

Haller, H. E. (2002). United States Patent No. 20020090268A1.

Hamada, E. (2007). Effects of microwave treatment on growth, photosynthetic pigments and some metabolites of wheat. Biologia Plantarum, 51(2), 343-345.

Joines, W. T. (2009). United States Patent No. US7601936B2.

Jolly, J. A., & Tate, R. L. (1971). Douglas- fir tree seed germination enhancement using microwave energy. Journal of Microwave Power, 6(2), 125-130.

Kalinin, L. G., Boshkova, I. L., Panchenko, G. I., & Kolomiichuk, S. G. (2005). Influence of low-frequency and microwave electromagnetic fields on seeds. Biophysics, 50(2), 334-337.

Kashyap, S. C., & Lewis, J. E. (1974). Microwave processing of tree seeds. Journal of Microwave Power, 9(2), 99-107.

Lal, R., & Reed, W. B. (1980). The effect of microwave-energy on germination and dormancy of wild oat seeds. Canadian Agricultural Engineering, 22(1), 85-88.

Mavrogianopoulos, G. N., Frangoudakis, A., & Pandelakis, J. (2000). Energy Efficient Soil Disinfestation by Microwaves. Journal of Agricultural Engineering Research, 75(2), 149-153

Menges, R. M., & Wayland, J. R. (1974). UHF electromagnetic energy for weed control in vegetables. Weed Science, 22(6), 584-590.

Monteiro, J. H., Mendiratta, S. K., & Capitão, A. (2008). Effect of microwave fields on the germination period and shoot growth rate of some seeds. Paper presented at the International Conference of Recent Advances in Microwave Theory and Applications, Jaipur, Rajasthan.

Nelson, S. O. (1996). A review and assessment of microwave energy for soil treatment to control pests. Transactions of the ASAE, 39(1), 281-289.

Nelson, S. O., Ballard, L. A. T., Stetson, L. E., & Buchwald, T. (1976). Increasing legume seed-germination by VHF and microwave dielectric heating. Transactions of the ASAE, 19(2), 369-371.

Nelson, S. O., & Stetson, L. E. (1985). Germination responses of selected plant species to RF electrical seed treatment. Transactions of the ASAE, 28(6), 2051-2058.
Petrov, I. Y., Moiseeva, T. V., & Morozova, E. V. (1991). A possibility of correction of vital processes in plant cell with microwave radiation. Paper presented at the 1991 IEEE International Symposium on Electromagnetic Compatibility, Cherry Hill, NJ, USA.
Rao, Y. V. S., Chakravarthy, N. V. K., & Panda, B. C. (1989). Effect of microwave irradiation on germination and initial growth of mustard seeds. Indian Journal of Agronomy, 34(3), 378-379.
Sartorato, I., Zanin, G., Baldoin, C., & De Zanche, C. (2006). Observations on the potential of microwaves for weed control. Weed Research, 46(1), 1-9.
Scialabba, A., & Tamburello, C. (2002). Microwave effects on germination and growth of radish (Raphanus sativus L.) seedlings. Acta Botanica Gallica, 149(2), 113-123.
Sera, B., Stranak, V., Sery, M., Tichy, M., & Spatenka, P. (2008). Germination of Chenopodium Album in Response to Microwave Plasma Treatment. Plasma Science and Technology, 10(4), 506-511.
Skiles, J. W. (2006). Plant response to microwaves at 2.45 GHz. Acta Astronautica, 58(5), 258-263. doi:DOI: 10.1016/j.actaastro.2005.12.007
Stephenson, M. M. P., Kushalappa, A. C., Raghavan, G. S. V., & Mather, D. E. (1996). Response surface models to describe the effects and phytotoxic thresholds of microwave treatments on barley seed germination and vigour. Seed Science and Technology, 24(1), 49-65.
Tran, V. N. (1979). Effects of Microwave Energy on the Strophiole, Seed Coat and Germination of Acacia Seeds. Australian Journal of Plant Physiology, 6(3), 277-287.
Vela-Múzquiz, R. (1984). Control of field weeds by microwave radiation. Paper presented at the II International Symposium on Soil Disinfestation.
Velazquez-Marti, B., & Gracia-Lopez, C. (2004). Thermal effects of microwave energy in agricultural soil radiation. International Journal of Infrared and Millimeter Waves, 25(7), 1109-1122.
Velazquez-Marti, B., Gracia-Lopez, C., & de la Puerta, R. (2008). Work conditions for microwave applicators designed to eliminate undesired vegetation in a field. Biosystems Engineering, 100(1), 31-37.
Velazquez-Marti, B., Gracia-Lopez, C., & Marzal-Domenech, A. (2006). Germination Inhibition of Undesirable Seed in the Soil using Microwave Radiation. Biosystems Engineering, 93(4), 365-373.
Vidmar, M. (2005). An improved microwave weed killer. Microwave Journal, 48(10), 116-126.
Wall, G. W. (2009). United States Patent No. US7560673B2.
Wayland, J., Merkle, M., Davis, F., Menges, R. M., & Robinson, R. (1975). Control of weeds with UHF electromagnetic fields. Unkrautbekämpfung mit elektromagnetischen UHF-Feldern., 15(1), 1-5.
Zanche, C. d., Amista, F., Baldoin, C., Beria, S., & Giubbolini, L. (2003). Design, construction and preliminary tests of a microwave prototype for weed control. Rivista di Ingegneria Agraria, 34(2), 31-38.

# 7 Applying Microwave Energy to Plants and the Soil

## 7.1 Introduction

Several researchers have used various radiating structures (antennae) in their pursuit of microwave weed control strategies. These structures have included: open ended wave-guides (Haller, 2002; Vidmar, 2005); horn antennae (Zanche, et al., 2003; Brodie, 2013a; 2013b); use of a sliding multi-mode cavity applicator (Joines, 2009); parabolic reflectors to focus the microwave energy; or leaky wave-guides. Most of these systems are designed to radiate microwave energy into space; consequently, the volume of soil exposed to microwave radiation depends on the field radiation pattern of the applicator and the natural attenuation of microwave energy in the soil.

Microwave field attenuation rates depend entirely on the dielectric properties of the soil. These properties change significantly with moisture content, with higher soil moisture resulting in higher dielectric properties (Figure 7.1) and therefore more surface energy reflection and field attenuation in the soil, compared with dry soil. Soil structure also affects the soil's response to electromagnetic fields, with clay soils having a dielectric response to a broader band of frequencies (Figure 7.1) than sand (Figure 7.2); however, moist sand has a higher dielectric loss factor (Figure 7.2) than clay (Figure 7.1), at the same moisture content. Despite this, at important ISM frequencies (i.e. 922 MHz and 2450 MHz), the dielectric constant of sand is higher than that of clay and the dielectric loss of sand is lower than that of clay at all moisture contents (Figure 7.3). This implies that at ISM frequencies, sand will heat more slowly than clay, due to more surface reflection and less absorption of the microwave field; however, there will be deeper heating in the sand than the clay.

## 7.2 Microwave Applicators

Applied microwave power is directly linked to the field strength of the microwave radiation; therefore, the heating rate due to exposure to microwave fields is also linked to the field strength. The rate of temperature rise, due to microwave irradiation, is:

$$\frac{dT}{dt} = \frac{2\pi f \varepsilon_0 \kappa'' (\tau E)^2}{\rho C} \tag{7.1}$$

Where T is temperature (°C), t is time (s), f is the microwave frequency (Hz), $\varepsilon_0$ is the permittivity of free space, κ" is the relative loss factor of the heated material, τ is the transmission coefficient of the material's surface, E is the electric field strength of the microwave radiation (V $m^{-1}$), ρ is the bulk density of the heated material (kg $m^{-3}$), and C is the specific heat capacity of the material (J °$C^{-1}$ $kg^{-1}$).

The field strength at the inter-facial surface of the heated material will depend on the geometry and operating characteristics of the microwave applicator (Metaxas & Meredith, 1983). Applicators may include radiating antennas in open space, wave-guides or traveling wave applicators, single-mode resonators and multi-mode resonators (Metaxas & Meredith, 1983). Radiating antennas are widely used in low power communication and high powered radar systems, but are not commonly used for microwave heating; however, dielectric antennas, which are inserted into other applicator systems can be used as impedance matching devices to reduce reflections from the heated material (Daian, Taube, & Shramkov, 2004).

Traveling wave applicators are variations on standard wave-guides (Metaxas & Meredith, 1983, pp. 104-129). Single mode resonators are wave-guides with short circuits or irises located at the null points in the standing wave created inside the chamber (Metaxas & Meredith, 1983, pp. 151-207). Multi-mode resonators are somewhat similar in arrangement, but simultaneously support many resonant modes within the chamber (Meredith, 1994; Metaxas & Meredith, 1983, pp. 130-150). These are commonly used in domestic microwave ovens.

### 7.2.1 Wave-Guide Applicators

In general, a wave-guide consists of a hollow metallic tube of uniform cross section; although variations on this principle can be designed for specific purposes (US Naval Air Systems Command, 1999). Common cross-section shapes for wave-guide are rectangular and circular. The microwave field distribution inside the wave-guide depends on the dimensions of the guide, the operating frequency of the source, and the mode of propagation of the wave within the guide (Cronin, 1995).

It is possible to simultaneously propagate several modes of electromagnetic waves within a wave-guide; however, this is not common. If the wavelength of the impressed signal is shorter than the cut-off wavelength for a given mode, then the wave will propagate through the guide with minimal attenuation (Cronin, 1995). If the wavelength of the impressed signal is longer than the cut-off wavelength, the wave will be attenuated to a negligible value in a relatively short distance (Cronin, 1995). If operation is above the cut-off wavelength, the wave is said to be *evanescent* (Ramo, Whinnery, & Van Duzer, 1965). Modes that have the same cut-off wavelengths but different field distributions are said to be *degenerate* (Ramo et al., 1965). The dominant mode in a wave-guide is the mode having the longest cut-off wavelength. For rectangular wave-guides this is the $TE_{10}$ mode (Ramo et al., 1965). The cut off frequency for a rectangular wave guide is defined by:

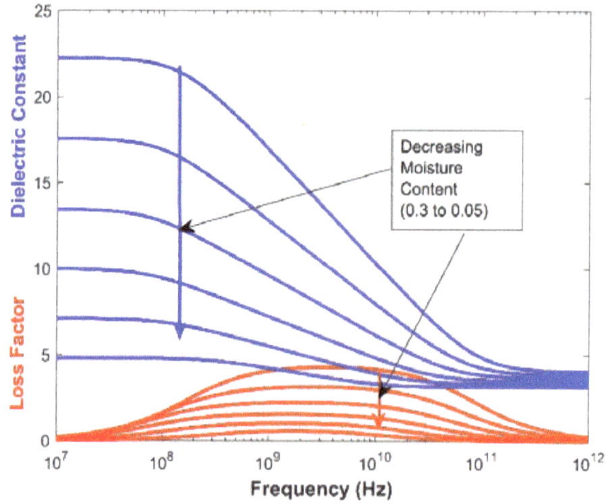


**Figure 7.1:** Dielectric properties of a clay based soil as a function of frequency and soil moisture (Data from: Wang, 1980).

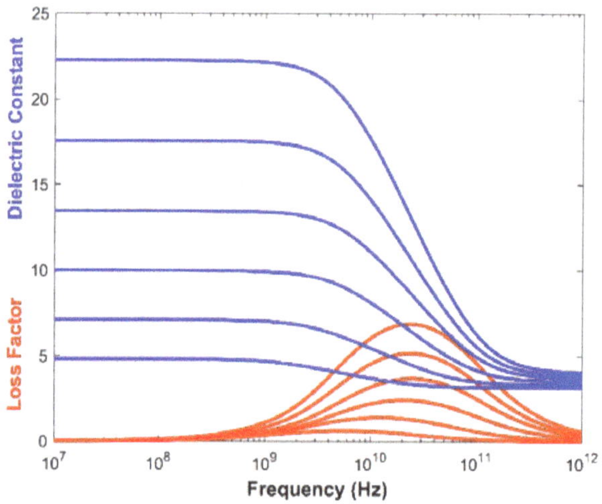


**Figure 7.2:** Dielectric properties of a sand based soil as a function of frequency and soil moisture (Data from: Wang, 1980).

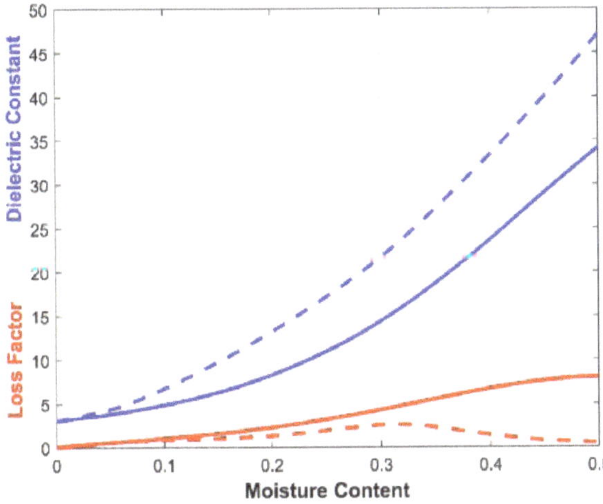


**Figure 7.3:** Dielectric properties of a sand (dotted) and clay (solid) as a function of soil moisture, at 2.45 GHz (Data from: Wang, 1980).

$$f_c = \frac{c}{2}\sqrt{\left(\frac{m}{a}\right)^2 + \left(\frac{n}{b}\right)^2} \tag{7.2}$$

Where a and b are the dimensions of the wave-guide (m), c is the speed of light (m $s^{-1}$), and m and n are mode integers for the wave form. The $TE_{10}$ mode has the lowest cut-off frequency for a rectangular wave guide and is usually chosen in practice. If the dimensions of the guide are chosen carefully, no other modes will propagate.

When a dielectric load, such as soil, is introduced into the wave-guide these field distributions are distorted by the reflections and internal attenuation associated with the material's dielectric properties.

In practical terms, wave-guide applicators allow conveyor belt processing techniques to be used, provided appropriately designed wave-guide chokes and feed tunnels are employed to prevent radiation from the applicator system into open space (Metaxas & Meredith, 1983, pp. 115).

### 7.2.2 Resonant Applicators

Single and multi-mode resonators are sections of wave-guide, with a short-circuit plate at the end. They usually require batch-processing techniques to heat materials. The electric field within multi-mode resonators, such as a microwave oven, is a very complex standing wave involving several modes. The resulting wave must satisfy Equation (7.3) (Metaxas & Meredith, 1983):

$$f_r = \frac{c}{2}\sqrt{\left(\frac{m}{a}\right)^2 + \left(\frac{n}{b}\right)^2 + \left(\frac{o}{d}\right)^2} \tag{7.3}$$

Where a, b, and d, are the dimensions of the cavity (m), c is the speed of light (m $s^{-1}$), and m, n, and o are mode integers for the wave form.

The resulting field distribution depends on the mode, or combination of modes, which establish within the cavity. A multi-mode resonator will simultaneously support both transverse electrical and transverse magnetic modes indicated as $TE_{lmn}$ and $TM_{lmn}$ (Meredith, 1994). The value of the index corresponds to periodicity (number of half waves) in the three principle axes of the cavity (Meredith, 1994).

The final field distribution in the cavity will be the sum of all the fields associated with the modes excited at any given frequency (Metaxas & Meredith, 1983). There is a spatially non-uniform field distribution within the cavity. Microwave heating occurs where peaks in the microwave fields occur. Little heating occurs at the nodes of the electric field, except in the case of small particle of ferrite based materials where eddy currents, created by the microwave's magnetic field, induce heating.

Because of this uneven field distribution, most domestic ovens employ a turntable to move the heated load through the electrical field to more evenly irradiate all exposed surfaces. Unfortunately, the resulting electric field strength is very difficult to determine theoretically as the inclusion of a dielectric material within a microwave oven effectively alters the field distribution (Metaxas & Meredith, 1983). Many authors (Perre & Turner, 1999; Unwin, 1999; Van Remmen, Ponne, Nijhuis, Bartels, & Herkhof, 1996; Zhao, Turner, & Torgovnikov, 1998) have adopted a variety of numerical techniques to predict the electric field strength within irradiated objects.

## 7.3 Antennas

Wave guides and resonant cavities require the treated material to be encapsulated by the applicator structure. This is not always practical, and is unrealistic for applications like microwave weed control. In these instances, an open structured applicator is required to heat the material. A common open structured applicator is an antenna.

Microwave heating of in situ materials, such as soil (Falciglia, Bonifacio, & Vagliasindi, 2016; Hur, Park, Kim, & Kim, 2013; Shibakova, 1975), asphalt (Salski et al., 2015; Tongsheng, 2016), and timber-in-service (Plaza et al., 2007) has been considered for some time. One of the challenges of in situ microwave heating is irradiation of the material without enclosing it in a cavity, such as an oven or wave guide (Metaxas & Meredith, 1983). Antennae, of various designs, are commonly used as microwave applicators in these cases (Brodie, 2013, 2017; Menges & Wayland, 1974; Spanu et al., 2016; Tongsheng, 2016; J. Wayland, Merkle, Davis, Menges, & Robinson, 1975; J. R.

Wayland, Davis, & Merkle, 1978). While short dipoles and other wire based antennae have found their place in medical applications, especially for cancer oblation, most in situ microwave treatments of larger objects use aperture antennae, such as open-ended wave guides, coaxial lines (Grosglik, Dikhtyar, & Jerby, 2002; E. Jerby, Aktushev, O., and Dikhtyar, V., 2005; E. Jerby, Dikhtyar, Aktushev, & Grosglick, 2002), reflector dishes, and horn antennae.

Antennas are the fundamental components of communication system, which uses free space as the transmitting medium; however, they will radiate energy into any medium. The theory behind antenna design, especially when used to illuminate a dielectric material, is not yet fully understood so the development of antennas usually involves trial and error to find a design that performs best in a certain system. Obviously, different systems transmit and receive different wavelengths, so the operational characteristics of the system are designed according to the properties of the antennas. The ideal transmitting antenna is one that will radiate all the power delivered to it in the desired direction and with the desired polarisation.

The antenna is the transition between a guiding device (transmission line, wave-guide) and the space before or around it. Its main purpose is to convert the energy of a guided wave into the energy of a propagated wave as efficiently as possible. It also acts as a focusing device to project the wave in the desired direction. Polarisation indicates the orientation of the electromagnetic wave's electric field, when radiated from the antenna (Connor, 1972). For example, a horizontally polarized antenna radiates energy with the electric field oriented into the horizontal direction, while a vertically polarised antenna radiates energy with the electric field oriented into the vertical direction (Connor, 1972).

Horn antennae have been widely used for microwave treatment of large objects (Diprose, Benson, & Willis, 1984; Joines, 2009; Mavrogianopoulos, Frangoudakis, & Pandelakis, 2000; Spanu et al., 2016; Tongsheng, 2016). Horn antennae are easily fabricated and, because they are radiating structures, can project microwave energy into the space occupied by the heated object.

## 7.4 Horn Antenna

For most microwave heating systems, horn antennae are open rectangular funnels, which fit onto a wave guide via a flange. The peak electric field strength ($E_0$) of a microwave propagating through a rectangular wave-guide in $TE_{10}$ mode is calculated by:

$$E_o = \sqrt{\frac{4\eta_o\overline{P}}{ab\sqrt{1-\left(\frac{n\lambda_o}{2a}\right)^2-\left(\frac{m\lambda_o}{2b}\right)^2}}} \tag{7.4}$$

Where $\eta_o$ is the electromagnetic wave impedance of free space, P is the mean power of the system (W), a and b are the cross-sectional dimensions of the wave guide (m), n and m are the electromagnetic mode numbers in the wave guide, and $\lambda_o$ is the wavelength of the electromagnetic wave in free space (m).

The microwave field expands as it traverses the taper of the antenna and field reflections occur at the mouth of the antenna due to the sudden change in impedance as the wave transits from the horn's aperture into open space. The transmission coefficient of the horn's aperture is given by:

$$\tau = \frac{2\sqrt{1-\left(\frac{n\lambda_o}{2a}\right)^2-\left(\frac{m\lambda_o}{2b}\right)^2}}{1+\sqrt{1-\left(\frac{n\lambda_o}{2a}\right)^2-\left(\frac{m\lambda_o}{2b}\right)^2}} \tag{7.5}$$

Based on the geometry of the horn antenna sown in Figure 7.4, the microwave field strength at a point P, with Cartesian coordinates of (x, y, z), in front of the aperture can be calculated using:

$$E_P = \frac{E_a}{4\pi}\int_{-B/2}^{B/2}\int_{-A/2}^{A/2}\cos\left(\frac{\pi}{A}x'\right)\frac{e^{-j\beta_o\left(\sqrt{(x-x')^2+(y-y')^2+z^2}+\sqrt{R_o^2+(x')^2}+\sqrt{R_o^2+(y')^2}\right)}}{\sqrt{(x-x')^2+(y-y')^2+z^2}}\cdot dx'\cdot dy' \tag{7.6}$$

Where $\beta_o$ is the electromagnetic wave number of free space ($m^{-1}$), x' and y' are coordinates in the aperture of the antenna (m), $R_o$ is the taper length of the antenna (m), and A and B are the cross-sectional dimensions of the antenna's aperture (m). Unfortunately, there is no simple solution for equation (7.6); however, several techniques can be used to approximate the field distribution. The integral can be evaluated: numerically, using Simpson's numerical approximation for the double integral; analytical, using an approximation and Fresnel Integrals (See Appendix to this chapter); or by simulation techniques, such as the Finite-Difference Time-Domain technique (A. Taflove, 1980, 1988, 1998; A. Taflove, Piket-May, & Hagness, 2000; Yee, 1966).

Figure 7.5 compares the various techniques for estimating the near field of a rectangular horn antenna. It is apparent, from Figure 7.5, that all the techniques provide similar results once the range from the antenna exceeds about 0.1 m. It is also apparent that the field strength quickly diminishes with distance from the aperture.

Because the near field of the horn antenna changes so rapidly with distance from the antenna's aperture, it is critical to establish a standardised microwave energy dose, to properly compare experiments and treatments. The simplest approach is

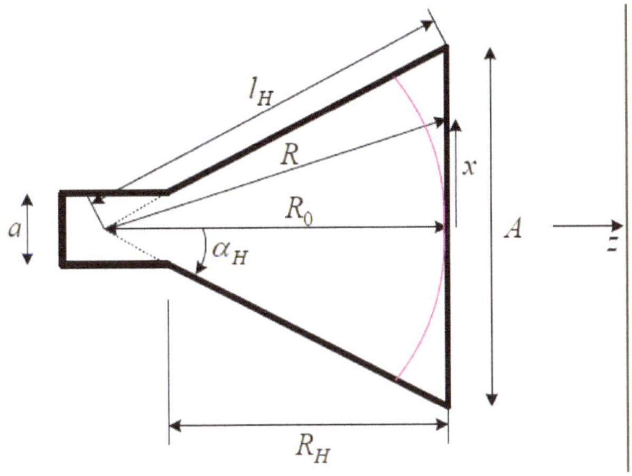


**Figure 7.4:** Geometry of a horn antenna showing a propagating wave front in the flare of the antenna (Nikolova, 2012).

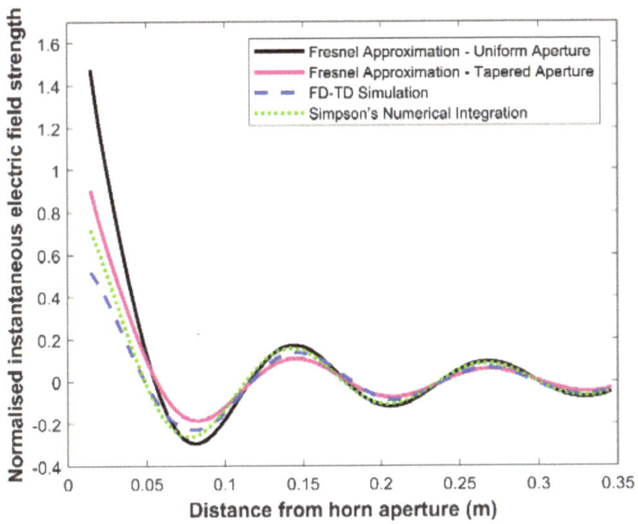


**Figure 7.5:** Comparison of estimated relative instantaneous field strength along the centre line in front of the horn antenna as a function of distance from the antenna's aperture plane.

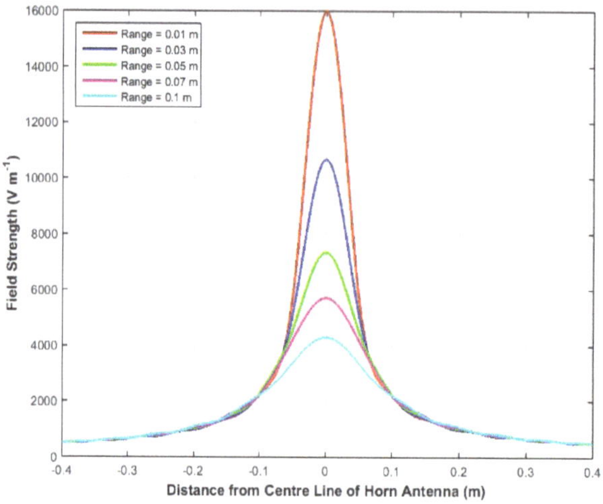


**Figure 7.6:** Estimated microwave field strength at ground level from a 2 kW microwave system as a function of height of the 110 mm by 55 mm antenna above the ground.

to use microwave field density at ground level as a standard. Figure 7.6 shows the estimated field strength as a function of range from the horn antenna and lateral distance from the centre line of a horn antenna. All weed and soil treatment doses in this book have been standardised using this approach, to allow ready comparisons across experiments. If the dose energy in literature is not clearly defined, it has been assumed that it is energy density at ground level.

The power density in a propagating field in front of the antenna is given by:

$$\overline{P}_p = \frac{E_p^2}{\eta_o} \tag{7.7}$$

This power density ($P_p$) at the point P, which lies anywhere in the ground plane, can be multiplied by the treatment time to calculate the applied microwave energy density at the soil surface.

## 7.5 Soil Temperature

It has been demonstrated (Brodie, 2008; Brodie, Hamilton, & Woodworth, 2007) that the temperature due to microwave heating in a semi-infinite solid, like soil, is given by:

$$T = \frac{n\omega\varepsilon_o\kappa''}{16k\alpha^2}\left(e^{4\gamma\alpha^2 t} - 1\right)\left[e^{-2\alpha z} + \left(\frac{h}{k} + 2\alpha\right)\cdot z \cdot e^{\frac{-z^2}{4\gamma t}}\right]\cdot E_p^2 \tag{7.8}$$

Where τ is the transmission coefficient of the solid's surface, γ is the joint heat and moisture transport coefficient (Henry, 1948), z is the distance into the soil (m), h is the convective surface heat transfer coefficient, k is the thermal conductivity of the soil, t is the microwave heating time (s), $E_p$ is defined in equation (7.6), and the microwave field attenuation in the soil is given by:

$$\alpha = \frac{2\pi f}{c}\sqrt{\frac{\kappa'}{2}\left(\sqrt{1+\left(\frac{\kappa''}{\kappa'}\right)^2}-1\right)} \tag{7.9}$$

Where c is the speed of light in free space, f is the frequency (Hz), κ' is the dielectric constant of the soil and κ" is the dielectric loss of the soil.
The transmission coefficient at the surface of the soil is defined by:

$$\tau = \frac{2\sqrt{\epsilon_1}}{\sqrt{\epsilon_1}+\sqrt{\epsilon_2}} \tag{7.10}$$

The effective volume of soil, which is being heated, is a semi-ellipsoid, the volume of which is:

$$V = \frac{2}{3}\cdot\frac{r^2}{\alpha} \tag{7.11}$$

where r is the effective beam width of the microwave fields.

The effective beam width will be the radius at which the microwave field strength reaches: $\frac{E_{peak}}{e}$; where $E_{peak}$ is the peak field strength on the centre line of the horn antenna, and e = 2.71828.

Therefore, the effective volume of heated soil depends on the range of the horn antenna from the soil surface, the soil texture and the moisture content of the soil. For example, the moist soil volume (MC = 0.2 on a dry mass basis) that the trailer mounted four by 2 kW prototype microwave system (Figure 7.7) will attempt to heat ranges between 0.1 and 1.0 litres, depending on the height of the horn antenna above the soil surface (range from 0.01 m to 0.1 m), although the temperature distribution within this volume will not be uniform (Figure 7.8). The estimated heating rate for a 110 mm by 55 mm horn antenna, operating at 2.45 GHz, at a range of 40 mm above the soil surface, is 1.7°C $kW^{-1}$ $s^{-1}$. The veracity of the model, outlined here, has been demonstrated experimentally (Brodie et al., 2007) on several occasions (Figure 7.9).

Figure 7.7: Trailer mounted four by 2 kW microwave weed killer.

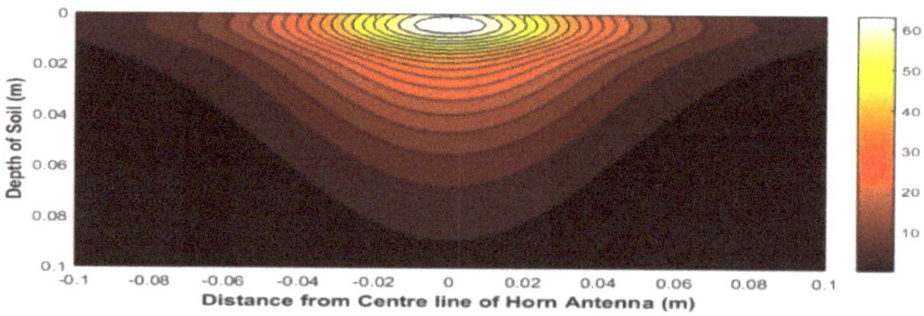


Figure 7.8: Calculated temperature change distribution in clay soil after 30 seconds of heating using the 2 kW microwave system and a horn antenna with an aperture of 110 mm by 55 mm at a range of 4 cm above the soil surface.

### 7.5.1 Using the Model to Understand Heat Distributions

This temperature model, proposed in equation (7.8), provides useful insight into the heating profile in the soil as parameters such as microwave treatment time, soil properties and antenna height are varied (Figure 7.10).

Having the antenna well above the soil surface provides a wider soil surface coverage but low intensity heating; however, having the antenna close to the soil surface provides a smaller treatment footprint but more intense and therefore faster heating. Unlike conventional heating, the maximum temperature occurs below the

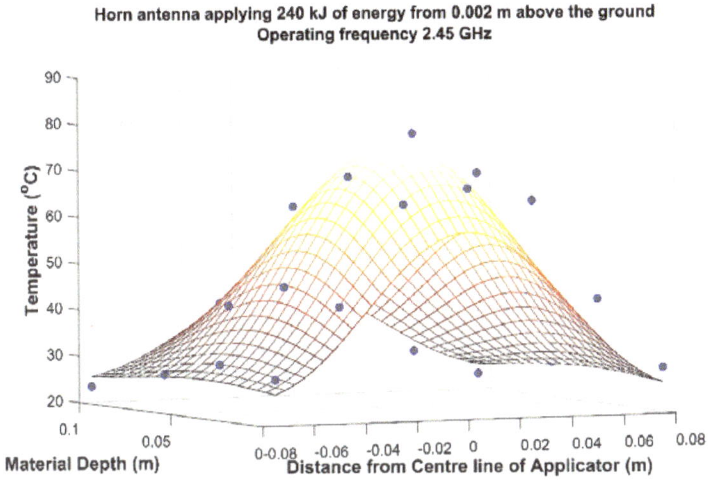


**Figure 7.9:** Comparison of temperature profile as predicted by equation (7.8), indicated by the mesh, and the measured temperature profile, indicated by the point data, after heating sand for 240 s using a 1 kW microwave oven prototype system with a 180 mm by 90 mm horn antenna.

soil surface rather than at the surface. This is a common feature of microwave heating (Brodie, 2008; Zielonka & Dolowy, 1998).

Soil moisture affects the temperature distribution during microwave treatment. This is mostly due to the increase in the dielectric properties of the soil as moisture increases. These increases in dielectric properties with increasing soil moisture reduce the penetration of microwave fields into the soil, but increase the heating rate of soil in the upper layers (Figure 7.11).

Soil texture also affects microwave heating (Figure 7.12). As suggested earlier, there is less heating in sandy soil than in clay soil, even when the soils are at the same moisture content; however, heating in the sandy soil will be deeper than in the clay soil. This is mostly due to the differences in dielectric properties of the soils at ISM frequencies.

Microwave frequency also affects the heating rate and distribution in the soil as well. Figure 7.13 shows the anticipated temperature profile in clay soil resulting from the two different ISM frequencies. The most important contribution to these differences in temperature profile are the larger volume of soil being heated at the lower frequency, due to the larger dimensions of the wave guide and antenna, and the differences in dielectric properties of the soil at the two frequencies. Using the lower frequency of 922 MHz (Australian ISM frequency) results in a heating of about 23 times the volume of soil, compared with using 2.45 GHz frequency. There is a consequent increase in energy requirements to achieve this increased volumetric heating; however, the energy requirements for 922 MHz is only 2.6 times higher than for 2.45 GHz, to

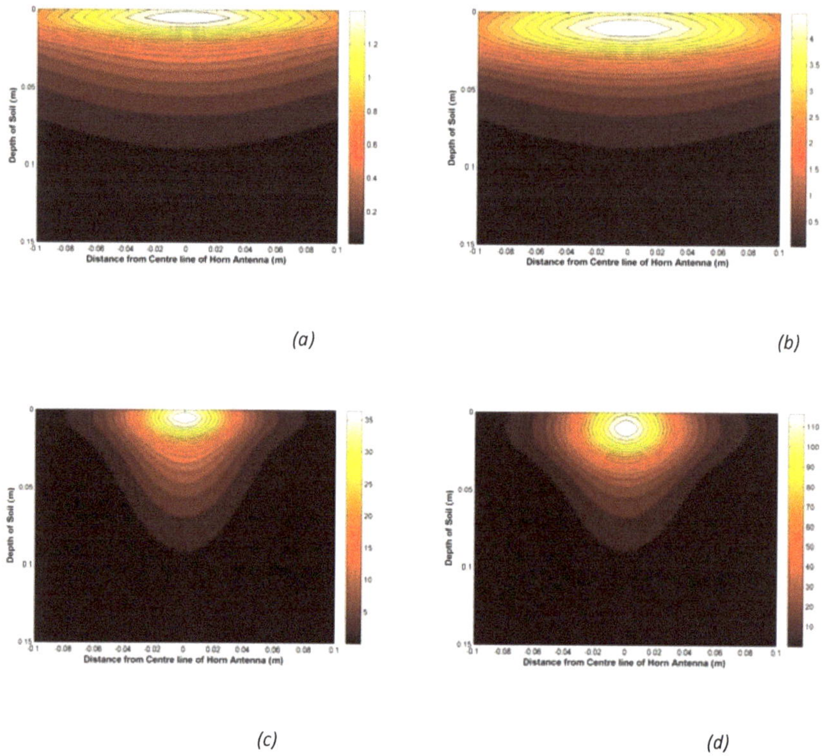


**Figure 7.10:** Estimated change in soil temperature from the 2 kW microwave system after (a) 10 seconds with antenna at 18 cm above the soil, (b) 30 seconds with antenna at 18 cm above the soil, (c) 10 seconds with antenna at 2 cm above the soil, and (d) 30 seconds with antenna at 2 cm above the soil.

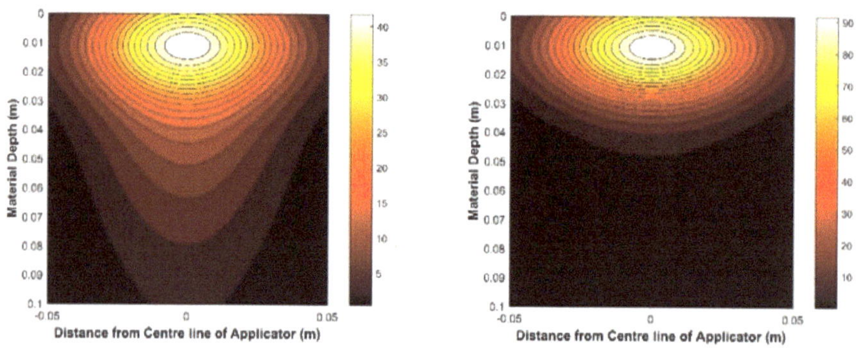


**Figure 7.11:** Comparison of temperature change profile after 60 seconds (at 2.45 GHz) as predicted by equation (7.8) in dry soil (10% MC) left and moist soil (35% MC) on the right.

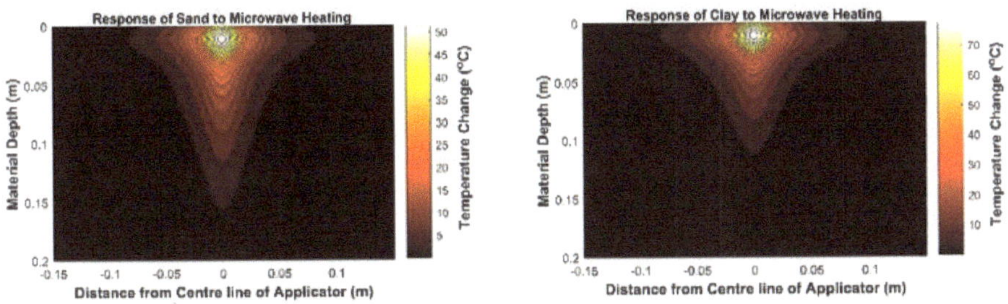


**Figure 7.12:** Comparison of temperature change profile after 90 seconds of microwave heating, at 2.45 GHz, in sandy soil (left), compared with clay soil (right).

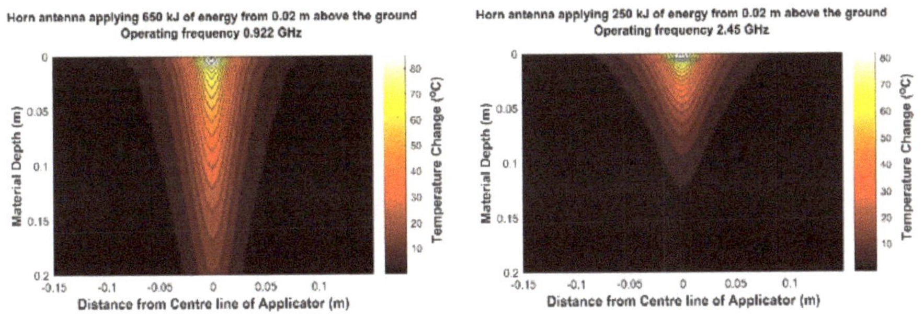


**Figure 7.13:** Comparison of temperature change in clay soil (left) treated using 922 MHz frequency and (right) treated using 2450 MHz frequency.

achieve the same maximum temperature. Therefore, adopting the lower frequency may provide some significant benefits for soil treatment on a larger scale.

## 7.6 Novel Design Considerations

Effective soil and weed treatment depends on field uniformity over a large area. Because of the electromagnetic field distribution in the aperture of a horn antenna, using an E-Plane flare, rather than a pyramidal horn or an H-Plane flare, provides a more uniform field across the flare of the antenna. If the antenna and mounting the systems is set up so that the E-plane of the antenna is perpendicular to the line of machine travel, this arrangement usually results in more uniform application of the microwave energy onto the soil and plants as the machine travels along.

Communication engineers use segmented apertures when the flare of the antenna becomes large. The segmentation of the antenna splits the field as it travels along the flare. In communications systems, segmentation lends itself to beam-forming and wider band-widths. As an applicator, segmentation provides more uniform field distribution across the aperture of the antenna. Segmentation can be achieved via an array of small antennae (Vidmar, 2005) or via internal separators being inserted in a single antenna.

Another interesting feature that is commonly used in communication antennae is corrugations on the walls of the antenna. These are used to better shape the field distributions in the aperture of the antenna. These corrugations introduce longitudinal field components (Connor, 1972; Georgieva, 2001) into the system, which create a traveling wave along the axis of the antenna. Travelling wave systems have wide application in particle accelerators (Tronc, 1987), because they lengthen the interaction between a microwave field and a material (usually a stream of charged particles), compared with a normal microwave applicator.

## 7.7 Conclusion

Horn antennae are commonly used to apply microwave energy to the soil. The resulting heating effect will depend on the geometry of the antenna, the soil moisture, the frequency of the microwave fields, the soil texture, the applied energy, and the height of the antenna above the ground.

## 7.8 References

Brodie, G. (2008). The influence of load geometry on temperature distribution during microwave heating. Transactions of the American Society of Agricultural and Biological Engineers, 51(4), 1401-1413.

Brodie, G. (2013). A Study of Microwave Heating in Plants using a Simple Horn Antenna as an Applicator. Paper presented at the 14th International Conference on Microwave and High Frequency Heating, Nottingham, UK.

Brodie, G. (2017). Microwave Heating Applicator Design. Paper presented at the 15th Australian Symposium on Antennas, Sydney.

Brodie, G., Hamilton, S., & Woodworth, J. (2007). An assessment of microwave soil pasteurization for killing seeds and weeds. Plant Protection Quarterly, 22(4), 143-149.

Connor, F. R. (1972). Antennas. London: Edward Arnold.

Cronin, N. J. (1995). Microwave and Optical Waveguides. Bristol: J W Arrowsmith Ltd.

Daian, M., Taube, A., & Shramkov, Y. (2004). Reducing the energy reflection from an applicator suitable for microwave wood processing. Paper presented at the Fourth World Congress on Microwave and Radio Frequency Appications, Austin.

Diprose, M. F., Benson, F. A., & Willis, A. J. (1984). The Effect of Externally Applied Electrostatic Fields, Microwave Radiation and Electric Currents on Plants and Other Organisms, with Special Reference to Weed Control. Botanical Review, 50(2), 171-223.

Falciglia, P. P., Bonifacio, A., & Vagliasindi, F. G. A. (2016). An Overview on Microwave Heating Application for Hydrocarboncontaminated Soil and Groundwater Remediation. Oil and Gas ResearchOil and Gas Research, 2(1), 1-6.

Georgieva, N. (2001). Modern Antennas in Wireless Telecommunications. Retrieved from http://www.ece.mcmaster.ca/faculty/georgieva/
Grosglik, U., Dikhtyar, V., & Jerby, E. (2002). Coupled Thermal-Electromagnetic Model for Microwave Drilling. Paper presented at the European Symposium on Numerical Methods in Electromagnetics, Toulouse, France.
Hur, J., Park, S.-W., Kim, M. C., & Kim, H. S. (2013). Enhanced binding of hydrophobic organic contaminants by microwave-assisted humification of soil organic matter. Chemosphere, 93, 2704-2710.
Jerby, E., Aktushev, O., and Dikhtyar, V. (2005). Theoretical analysis of the microwave-drill near-field localized heating effect. Journal of Applied Physics, 97(3), 034909-034901 - 034909-034907.
Jerby, E., Dikhtyar, V., Aktushev, O., & Grosglick, U. (2002). The microwave drill. Science, 298(5593), 587-589.
Joines, W. T. (2009). United States Patent No. US7601936B2.
Mavrogianopoulos, G. N., Frangoudakis, A., & Pandelakis, J. (2000). Energy Efficient Soil Disinfestation by Microwaves. Journal of Agricultural Engineering Research, 75(2), 149-153
Menges, R. M., & Wayland, J. R. (1974). UHF electromagnetic energy for weed control in vegetables. Weed Science, 22(6), 584-590.
Meredith, R. J. (1994). A three axis model of the mode structure of multimode cavities. The Journal of Microwave Power and Electromagnetic Energy, 29(1), 31-44.
Metaxas, A. C., & Meredith, R. J. (1983). Industrial Microwave Heating. London: Peter Peregrinus.
Nikolova, N. K. (2012). Modern Antennas in Wireless Telecommunications. Retrieved from http://www.ece.mcmaster.ca/faculty/nikolova/antennas.htm
Perre, P., & Turner, I. W. (1999). The use of numerical simulation as a cognitive tool for studying the microwave drying of softwood in an over-sized wave-guide. Wood Science and Technology, 33(6), 445-464.
Plaza, P. J., Zona, A. T., Sanshis, R., Balbastre, J. V., Martinez, A., Munoz, E. M., et al. (2007). Microwave disinfestation of bulk timber. Journal of Microwave Power and Electromagnetic Energy, 41(3), 21-36.
Ramo, S., Whinnery, J. R., & Van Duzer, T. (1965). Fields and Waves in Communication Electronics: John Wiley.
Salski, B., Olszewska-Placha, M., Rudnicki, J., Gwarek, W., Karpisz, T., & Reszewicz, S. (2015, 17-22 May 2015). Design of an applicator for microwave-assisted bituminous surface thermal bonding. Paper presented at the Microwave Symposium (IMS), 2015 IEEE MTT-S International.
Shibakova, V. S. (1975). On the possibility of using microwave energy for soil stabilization. Bulletin of the International Association of Engineering Geology - Bulletin de l'Association Internationale de Géologie de l'Ingénieur, 12(1), 89-91. doi:10.1007/bf02635436
Spanu, M., Fanti, A., Bruno Lodi, M., Casu, S., Desogus, F., Bisceglia, B., & Mazzarella, G. (2016). Microwaves disinfection of farmland. Journal of Electromagnetic Waves and Applications, 1-9. doi:10.1080/09205071.2016.1249802
Taflove, A. (1980). Applications of the Finite-Difference Time-Domain method to sinusoidal steady-state electrimagnetic penetration problems. IEEE Transactions on Electromagnetic Compatibility, EMC-22(3), 191-202.
Taflove, A. (1988). Review of the formulation and applications of the finite-difference time-domain method for numerical modeling of electromagnetic wave interactions with arbitrary structures. Wave Motion, 10(6), 547-582.
Taflove, A. (1998). Computational Electrodynamics: The Finite-Difference Time-Domain Method. Boston: Artech House.
Taflove, A., Piket-May, M. J., & Hagness, S. C. (2000). FDTD-how complex a problem can we solve? Paper presented at the Antennas and Propagation Society International Symposium, 2000.

Tongsheng, S. (2016). Directivity factor of coupled antennas for microwave heating of asphalt mixture. International Journal of Applied Electromagnetics & Mechanics, 50(4), 647-663. doi:10.3233/JAE-150153
Tronc, D. (1987). Travelling-wave particle accelerator including an energy recirculation device, CGR MEV.
Unwin, C. E. (1999). Computer Modelling of Temperature Distributions in Foods during Microwave Heating. (Master of Science), University of Aberdeen, Aberdeen.
US Naval Air Systems Command. (1999). Electronic Warfare and Radar Systems Engineering Handbook: Avionics Department, Naval Air Systems Command, US Navy.
Van Remmen, H. H. J., Ponne, C. T., Nijhuis, H. H., Bartels, P. V., & Herkhof, P. J. A. M. (1996). Microwave Heating Distribution in Slabs, Spheres and Cylinders with Relation to Food Processing. Journal of Food Science, 61(6), 1105-1113.
Vidmar, M. (2005). An improved microwave weed killer. Microwave Journal, 48(10), 116-126.
Wayland, J., Merkle, M., Davis, F., Menges, R. M., & Robinson, R. (1975). Control of weeds with UHF electromagnetic fields. Unkrautbekämpfung mit elektromagnetischen UHF-Feldern., 15(1), 1-5.
Wayland, J. R., Davis, F. S., & Merkle, M. G. (1978). United States Patent No.: C. S. Phytox Corporation, Texas.
Yee, K. S. (1966). Numerical solution of initial boundary value problems involving Maxwell's equations in isotropic media. IEEE Transactions on Antennas and Propagation, 14(3), 302-307.
Zhao, H., Turner, I. W., & Torgovnikov, G. (1998). An experimental and numerical investigation of the microwave heating of wood. The Journal of Microwave Power and Electromagnetic Energy, 33(2), 121-133.
Zielonka, P., & Dolowy, K. (1998). Microwave Drying of Spruce: Moisture Content, Temperature and Heat Energy Distribution. Forest Products Journal, 48(6), 77-80.

## 7.9 Appendix A - Derivation of Near Field from a Uniformly Illuminated Rectangular Aperture

In this case the field along the centre line of the antenna will be approximated by:

$$E_P = \frac{zE_o}{j\lambda} \int_{-B/2}^{B/2} \int_{-A/2}^{A/2} \frac{e^{-j\beta_o \sqrt{(x-x')^2+(y-y')^2+z^2}}}{(x-x')^2+(y-y')^2+z^2} \cdot dx' \cdot dy' \tag{7.12}$$

Let $\rho^2 = (x-x')^2 + (y-y')^2$ and $r = \sqrt{z^2+\rho^2} = z\sqrt{1+\frac{\rho^2}{z^2}}$

If $\frac{\rho^2}{z^2} < 1.0$ then $z\sqrt{1+\frac{\rho^2}{z^2}} \approx z\left(1+\frac{\rho^2}{2z^2}\right)$

Substituting into (C1) and ignoring the $(x-x')^2 + (y-y')^2$ in the denominator of the integrand yields:

$$E_P \approx \frac{E_o}{j\lambda z} \int_{-B/2}^{B/2} \int_{-A/2}^{A/2} e^{-j\beta_o z\left(1+\frac{(x-x')^2+(y-y')^2}{2z}\right)} \cdot dx' \cdot dy' \tag{7.13}$$

Now $e^{-j\theta} = Cos(-\theta) + jSin(-\theta)$ therefore:

$$E_P \approx \frac{E_o e^{-j\beta_o z}}{j\lambda z}\left\{\begin{array}{l}\int_{-B/2}^{B/2}\int_{-A/2}^{A/2} Cos\left(-\beta_o \frac{(x-x')^2+(y-y')^2}{2}\right)\cdot dx'\cdot dy' + \\ j\int_{-B/2}^{B/2}\int_{-A/2}^{A/2} Sin\left(-\beta_o \frac{(x-x')^2+(y-y')^2}{2}\right)\cdot dx'\cdot dy'\end{array}\right\} \quad (7.14)$$

Considering only the real part of this equation yields:

$$E_P \approx \frac{\pi E_o e^{-j\beta_o z}}{\beta_o \lambda z}\left\{C\left[\sqrt{\frac{\beta_o}{\pi}}(x-x')\right]S\left[-\sqrt{\frac{\beta_o}{\pi}}(y-y')\right]+S\left[\sqrt{\frac{\beta_o}{\pi}}(x-x')\right]C\left[-\sqrt{\frac{\beta_o}{\pi}}(y-y')\right]\right\}\Bigg|_{x'=-A/2}^{A/2}\Bigg|_{y'=-B/2}^{B/2} \quad (7.15)$$

$$E_P \approx \frac{j\pi E_o e^{-j\beta_o z}}{2\beta_o \lambda z}\left\{\begin{array}{l} S\left[-\sqrt{\frac{\beta_o}{\pi}}\left(\frac{A}{2}+x\right)\right]\left\{S\left[\sqrt{\frac{\beta_o}{\pi}}\left(\frac{B}{2}-y\right)\right]-S\left[-\sqrt{\frac{\beta_o}{\pi}}\left(\frac{B}{2}+y\right)\right]\right\}- \\ C\left[-\sqrt{\frac{\beta_o}{\pi}}\left(\frac{A}{2}+x\right)\right]\left\{C\left[\sqrt{\frac{\beta_o}{\pi}}\left(\frac{B}{2}-y\right)\right]-C\left[-\sqrt{\frac{\beta_o}{\pi}}\left(\frac{B}{2}+y\right)\right]\right\}+ \\ C\left[\sqrt{\frac{\beta_o}{\pi}}\left(\frac{A}{2}-x\right)\right]\left\{C\left[\sqrt{\frac{\beta_o}{\pi}}\left(\frac{B}{2}-y\right)\right]-C\left[-\sqrt{\frac{\beta_o}{\pi}}\left(\frac{B}{2}+y\right)\right]\right\}- \\ S\left[\sqrt{\frac{\beta_o}{\pi}}\left(\frac{A}{2}-x\right)\right]\left\{S\left[\sqrt{\frac{\beta_o}{\pi}}\left(\frac{B}{2}-y\right)\right]-S\left[-\sqrt{\frac{\beta_o}{\pi}}\left(\frac{B}{2}+y\right)\right]\right\} \end{array}\right\} \quad (7.16)$$

## 7.10 Appendix B - Derivation of Near Field from a Tapered Rectangular Aperture

The field along the centre line of the antenna will be approximated by:

$$E_p \approx \frac{E_o e^{j\omega t}}{\lambda}\int_{-\frac{B}{2}}^{\frac{B}{2}}\int_{-\frac{A}{2}}^{\frac{A}{2}} Cos\left[\frac{\pi x'}{A}\right]\cdot\frac{e^{-j\beta_o\sqrt{(x-x')^2+(y-y')^2+z^2}}}{\sqrt{(x-x')^2+(y-y')^2+z^2}}\cdot dx'\cdot dy' \quad (7.17)$$

Let $\rho^2 = (x-x')^2 + (y-y')^2$, if $\frac{\rho^2}{z^2} < 1.0$ then $z\sqrt{1+\frac{\rho^2}{z^2}} \approx z\left(1+\frac{\rho^2}{2z^2}\right)$

Substituting into (1) and ignoring the $(x-x')^2+(y-y')^2$ in the denominator of the integrand yields:

$$E_p \approx \frac{E_o e^{j\omega t}}{\lambda z} \int_{-\frac{B}{2}}^{\frac{B}{2}} \int_{-\frac{A}{2}}^{\frac{A}{2}} Cos\left[\frac{\pi x'}{A}\right] \cdot e^{-j\beta_o z - j\beta_o \frac{(x-x')^2+(y-y')^2}{2z}} \cdot dx' \cdot dy' \qquad (7.18)$$

Now $e^{-j\theta} = Cos(-\theta) + jSin(-\theta)$ therefore:

$$E_p \approx \frac{E_o e^{j(\omega t-\beta_o z)}}{\lambda z} \left\{ \begin{array}{l} \int_{-\frac{B}{2}}^{\frac{B}{2}} \int_{-\frac{A}{2}}^{\frac{A}{2}} Cos\left[\frac{\pi x'}{A}\right] \cdot Cos\left[-\beta_o \frac{(x-x')^2+(y-y')^2}{2}\right] \cdot dx' \cdot dy' + \\ j\int_{-\frac{B}{2}}^{\frac{B}{2}} \int_{-\frac{A}{2}}^{\frac{A}{2}} Cos\left[\frac{\pi x'}{A}\right] \cdot Sin\left[-\beta_o \frac{(x-x')^2+(y-y')^2}{2}\right] \cdot dx' \cdot dy' \end{array} \right\} \qquad (7.19)$$

Considering only the real part of this equation yields:

$$E_p \approx \frac{E_o e^{j(\omega t-\beta_o z)}}{\lambda z} \int_{-\frac{B}{2}}^{\frac{B}{2}} \int_{-\frac{A}{2}}^{\frac{A}{2}} Cos\left[\frac{\pi x'}{A}\right] \cdot Cos\left[-\beta_o \frac{(x-x')^2+(y-y')^2}{2}\right] \cdot dx' \cdot dy' \qquad (7.20)$$

From trigonometry:

$$E_p \approx \frac{E_o e^{j(\omega t-\beta_o z)}}{2\lambda z} \int_{-\frac{B}{2}}^{\frac{B}{2}} \int_{-\frac{A}{2}}^{\frac{A}{2}} \left\{Cos\left[-\beta_o \frac{(x-x')^2+(y-y')^2}{2} + \frac{\pi x'}{A}\right] + Cos\left[-\beta_o \frac{(x-x')^2+(y-y')^2}{2} - \frac{\pi x'}{A}\right]\right\} \cdot dy' \qquad (7.21)$$

Evaluating the first integral yields:

$$E_p \approx \frac{E_o e^{j(\omega t-\beta_o z)}}{2\lambda z} \int_{-\frac{B}{2}}^{\frac{B}{2}} \sqrt{\frac{\pi}{\beta_o}} \cos\left[\frac{\beta_o\left(x^2-(y-y')^2\right)}{2} - \frac{(\pi+A\beta_o x)^2}{2A^2\beta_o}\right] \cdot C\left[-2\sqrt{\frac{1}{\pi\beta_o}}\left(\frac{(\pi+A\beta_o x)}{2A} - \frac{\beta_o x'}{2}\right)\right] + \sqrt{\frac{\pi}{\beta_o}} \sin\left[\frac{\beta_o\left(x^2-(y-y')^2\right)}{2} - \frac{(\pi+A\beta_o x)^2}{2A^2\beta_o}\right] \cdot S\left[-2\sqrt{\frac{1}{\pi\beta_o}}\left(\frac{(\pi+A\beta_o x)}{2A} - \frac{\beta_o x'}{2}\right)\right] + \sqrt{\frac{\pi}{\beta_o}} \cos\left[\frac{\beta_o\left(x^2-(y-y')^2\right)}{2} - \frac{(\pi-A\beta_o x)^2}{2A^2\beta_o}\right] \cdot C\left[2\sqrt{\frac{1}{\pi\beta_o}}\left(\frac{(\pi-A\beta_o x)}{2A} + \frac{\beta_o x'}{2}\right)\right] + \sqrt{\frac{\pi}{\beta_o}} \sin\left[\frac{\beta_o\left(x^2-(y-y')^2\right)}{2} - \frac{(\pi-A\beta_o x)^2}{2A^2\beta_o}\right] \cdot S\left[2\sqrt{\frac{1}{\pi\beta_o}}\left(\frac{(\pi-A\beta_o x)}{2A} + \frac{\beta_o x'}{2}\right)\right] \cdot dy' \Bigg|_{x'=-\frac{A}{2}}^{\frac{A}{2}} \qquad (7.22)$$

Evaluating the second integral yields:

$$E_p \approx \frac{\pi E_o e^{j(\omega t-\beta_o z)}}{2\lambda\beta_o z}\cdot C\left[-\sqrt{\frac{1}{\pi\beta_o}}\left(\frac{(\pi+A\beta_o x)}{A}-\beta_o x'\right)\right]\cdot\left\{C\left[\sqrt{\frac{\beta_o}{\pi}}(y'-y)\right]\cos\left[\frac{\pi^2+2\pi A\beta_o x}{2A^2\beta_o}\right]+ S\left[\sqrt{\frac{\beta_o}{\pi}}(y'-y)\right]\sin\left[\frac{\pi^2+2\pi A\beta_o x}{2A^2\beta_o}\right]\right\}-S\left[-\sqrt{\frac{1}{\pi\beta_o}}\left(\frac{(\pi+A\beta_o x)}{A}-\beta_o x'\right)\right]\cdot \left\{S\left[\sqrt{\frac{\beta_o}{\pi}}(y'-y)\right]\cos\left[\frac{\pi^2+2\pi A\beta_o x}{2A^2\beta_o}\right]-C\left[\sqrt{\frac{\beta_o}{\pi}}(y'-y)\right]\sin\left[\frac{\pi^2+2\pi A\beta_o x}{2A^2\beta_o}\right]\right\}+C\left[\sqrt{\frac{1}{\pi\beta_o}}\left(\frac{(\pi-A\beta_o x)}{A}+\beta_o x'\right)\right]\cdot\left\{C\left[\sqrt{\frac{\beta_o}{\pi}}(y'-y)\right]\cos\left[\frac{\pi^2-2\pi A\beta_o x}{2A^2\beta_o}\right]+S\left[\sqrt{\frac{\beta_o}{\pi}}(y'-y)\right]\sin\left[\frac{\pi^2-2\pi A\beta_o x}{2A^2\beta_o}\right]\right\}- S\left[\sqrt{\frac{1}{\pi\beta_o}}\left(\frac{(\pi-A\beta_o x)}{A}+\beta_o x'\right)\right]\cdot \left\{S\left[\sqrt{\frac{\beta_o}{\pi}}(y'-y)\right]\cos\left[\frac{\pi^2-2\pi A\beta_o x}{2A^2\beta_o}\right]-C\left[\sqrt{\frac{\beta_o}{\pi}}(y'-y)\right]\sin\left[\frac{\pi^2-2\pi A\beta_o x}{2A^2\beta_o}\right]\right\}\Bigg|_{y'=-\frac{B}{2}}^{\frac{B}{2}}\Bigg|_{x'=-\frac{A}{2}}^{\frac{A}{2}} \tag{7.23}$$

This leads to:

$$E_p \approx \frac{\pi E_o e^{j(\omega t-\beta_o z)}}{2\lambda\beta_o z}\langle (CxNU-CxNL)\cdot\{(CyU-CyL)CosP+(SyU-SyL)SinP\}-(SxNU-SxNL)\cdot\{(SyU-SyL)CosP-(CyU-CyL)SinP\}+(CxPU-CxPL)\cdot\{(CyU-CyL)CosN+(SyU-SyL)SinN\}-(SxPU-SxPL)\cdot\{(SyU-SyL)CosN-(CyU-CyL)SinN\}\rangle \tag{7.24}$$

Where:

$CxNU = C\left[-\sqrt{\frac{1}{\pi\beta_o}}\left(\frac{(\pi+A\beta_o x)}{A}-\beta_o\frac{A}{2}\right)\right]$, $CxNL = C\left[-\sqrt{\frac{1}{\pi\beta_o}}\left(\frac{(\pi+A\beta_o x)}{A}+\beta_o\frac{A}{2}\right)\right]$ ,$CyU = C\left[\sqrt{\frac{\beta_o}{\pi}}\left(\frac{B}{2}-y\right)\right]$, $CyL = C\left[-\sqrt{\frac{\beta_o}{\pi}}\left(\frac{B}{2}+y\right)\right]$, $CosP = \cos\left[\frac{\pi^2+2\pi A\beta_o x}{2A^2\beta_o}\right]$, $SyU = S\left[\sqrt{\frac{\beta_o}{\pi}}\left(\frac{B}{2}-y\right)\right]$, $SyL = S\left[-\sqrt{\frac{\beta_o}{\pi}}\left(\frac{B}{2}+y\right)\right]$, $SinP = \sin\left[\frac{\pi^2+2\pi A\beta_o x}{2A^2\beta_o}\right]$, $SxNU = S\left[-\sqrt{\frac{1}{\pi\beta_o}}\left(\frac{(\pi+A\beta_o x)}{A}-\beta_o\frac{A}{2}\right)\right]$, $SxNL = S\left[-\sqrt{\frac{1}{\pi\beta_o}}\left(\frac{(\pi+A\beta_o x)}{A}+\beta_o\frac{A}{2}\right)\right]$, $CosN = \cos\left[\frac{\pi^2-2\pi A\beta_o x}{2A^2\beta_o}\right]$, $SinN = \sin\left[\frac{\pi^2-2\pi A\beta_o x}{2A^2\beta_o}\right]$,

$CxPU = C\left[\sqrt{\frac{1}{\pi\beta_o}}\left(\frac{(\pi-A\beta_o x)}{A}+\beta_o\frac{A}{2}\right)\right]$, $CxPL = C\left[\sqrt{\frac{1}{\pi\beta_o}}\left(\frac{(\pi-A\beta_o x)}{A}-\beta_o\frac{A}{2}\right)\right]$, $SxPU = S\left[\sqrt{\frac{1}{\pi\beta_o}}\left(\frac{(\pi-A\beta_o x)}{A}+\beta_o\frac{A}{2}\right)\right]$, and $SxPL = S\left[\sqrt{\frac{1}{\pi\beta_o}}\left(\frac{(\pi-A\beta_o x)}{A}-\beta_o\frac{A}{2}\right)\right]$.

# 8 The Potential of Microwave Treatment to Kill Weed Plants

## 8.1 Introduction

The effect of radio frequency and microwave radiation on plants has interested researchers for some time, with several experiments dating to the 1920's (Ark & Parry, 1940). Magone (1996) studied duckweed (*Spirodela polyrhiea*) grown in flasks some 2 km from a radio station transmitter. The frequency and intensity of the radiation used was 156–162 MHz and 0.1–1.8 μW $cm^{-2}$. Generally, the vegetative reproduction rate of plants that were exposed to the electromagnetic radiation for between 24 hours and 88 hours was accelerated by between 105% and 195%, compared with the control plants, during the first 20 days after exposure. Exposure of plants that were just beginning formation exhibited slightly reduced vegetative growth rate. This phenomenon has also been observed in yeast exposed to 40 to 60 GHz microwave fields by Grundler, Keilmann, and Fröhlich (1977) and by Horikoshi, Hasegawa, and Suzuki (2017), who showed that plant growth and the onset of reproductive growth can be significantly enhanced by very short exposure to very low intensity, 2.45 GHz microwave irradiation.

It is likely that this increased growth and particularly the onset of reproductive growth is a stress response in the plant. Controlled deficit irrigation (Zapata-Sierra & Manzano-Agugliaro, 2017) is a more conventional strategy, which also achieves a reproductive response in plants due to the release of stress hormones. Controlled deficit irrigation allows parts of the root zone to dry out while other parts of the root zone are watered. Watering is usually alternated around the root zone, so that the roots in the deficit area are not compromised.

Application of very low level electromagnetic fields seems to have a more beneficial effect on the plants than simply allowing heat to dry the root zones. While studying the influence of microwave radiation on the plant cell membrane transport system activity, Petrov, Moiseeva, and Morozova (1991) found that membrane potential difference change deviation was the opposite to that induced by traditional heating, resulting in a positive response in plant growth. Although this is a fascinating phenomenon, it is not the focus of this work.

Skiles (2006) conducted an experiment where alfalfa was exposed to continuous microwave energy at 2.45GHz with intensities of 0.5 – 1.2 mW $cm^{-2}$. After 7 weeks, the plants were harvested and fresh weight and dry weight were measured. There was no difference between the control and the microwave treated plants.

Wayland, Merkle, Davis, Menges, and Robinson (1975) demonstrated that microwave energy, with an energy density of 35 J $cm^{-2}$ or more significantly damaged or killed growing plants. Values of 77, 800 and 1600 J $cm^{-2}$ have all been quoted as the minimum energy necessary for effective weed control (Hightower, Burdette, &

Burns, 1974; Wayland et al., 1975). Microwave radiation has several advantages when used to treat weeds. These include rapid penetration to all parts of the plant, without leaving any residue after application. The ability to kill herbicide resistant plants without disturbing the soil or adjacent crop plants, because microwave energy can be specifically focused on the target plants (Diprose, Benson, & Willis, 1984). Microwave radiation is not affected by wind or rain. This extends the periods of application, compared to conventional spraying methods and has the potential to overcome crop yield losses due to treatment timeliness issues.

It is apparent that very small doses of electromagnetic energy can induce stress hormone responses in plants, which lead to improved reproductive production, while larger doses cause damage to plants. In weed management, the objective is to selectively damage unwanted plants.

## 8.2 Early Experimental Prototypes

Early work by Brodie and colleagues (Brodie, Botta, & Woodworth, 2007; Brodie et al., 2009) used a modified microwave oven, with a wave guide and horn antenna feeding through and out of the oven cavity to apply energy to plants and the soil. The microwave oven (Sanyo Electric Co., 800 W, size 46 × 31 × 28 cm) used a rectangular (86 ×43 mm internal dimensions) wave-guide to channel the microwave energy from the oven's 2.4 GHz magnetron to a pyramidal horn antenna outside of the oven (Figure 8.1). The horn (aperture dimensions of 180 × 90 mm and a length of 180 mm) was attached to the wave-guide via a 90° elbow. The prototype was used in many plant and soil heating experiments between 2007 and about 2010.

The original prototype, shown in Figure 8.1, was difficult to work with in field conditions, so a more flexible arrangement was devised (Figure 8.2). This system is more compact, with a housing and wave guide launcher that holds a 900 to 1200 W magnetron. This prototype is still wired into a microwave oven, replacing the connection to the oven's magnetron; therefore, the oven's voltage doubler circuit and electronic controls can still control the activity of the prototype's magnetron. The magnetron, located in the housing above the horn antenna, is cooled by an electric fan in the wall of the housing. This prototype design is still being used for experimental work.

Both early prototypes were useful for static experiments on pots and in the field; however, mobility is needed to test microwave weed control in field conditions. When funding became available, a trailer mounted microwave prototype was developed (Figure 8.3). This system has four fixed output 2 kW microwave generators operating at 2.45 GHz. It uses switched-mode power supplies to operate the water-cooled magnetron heads. The magnetron heads can be independently operated from a control panel in the side of the switching cabinet at the front of the trailer. The trailer system has been

**Figure 8.1:** Prototype system, fed from the magnetron of a 800 W microwave oven (Source: Bebawi et al., 2007).

**Figure 8.2:** Rendering of the second prototype system, with a 900 W magnetron. The magnetron is wired into the voltage doubler circuit of a normal microwave oven via a 2 m long cable.

**Figure 8.3:** Prototype four by 2 kW microwave weed killer (above) with details of the control panel (below).

used for both static experiments on potted specimens and for moving experiments in field conditions.

### 8.2.1 Static Experiments on Plants

Potted plants have been treated with all the microwave prototypes. Experiments have been conducted for the following species: annual ryegrass (*Lolium rigidum*); barley grass (*Hordeum vulgare* L.); barnyard grass (*Echinochloa crus-galli*); fleabane (*Conyza bonariensis* L.); marshmallow (*Malva parviflora* L.); prickly paddy melon (*Cucumis myriocarpus*); and wild oat (*Avena fatua*) (Brodie, Hamilton, & Woodworth, 2007; Brodie & Hollins, 2015; Brodie, Ryan, & Lancaster, 2012). In each case, both pot trials and in-situ field experiments were conducted for each species. Because of the potential hazard of directly measuring the microwave field strength in the aperture of the horn antenna, applied microwave energy, at ground level was determined using the models developed in Chapter 7.

In most cases the dose response for plant survival, as a function of applied microwave energy (Ψ), was described by:

$$S = a \bullet erfc[b \bullet (\Psi - c)] \tag{8.1}$$

Where S is the survival fraction of the treated population, Ψ is the microwave energy density at ground level (J $cm^{-2}$), *erfc(x)* is the Gaussian complementary error function of x, and a, b, and c are constants to be determined by regression for each experiment and species.

Equation (8.1) assumes that the individual plant responses to microwave irradiation are normally distributed, because the Gaussian Error functions represent the cumulative response of the population (i.e.: $erfc(z) = \frac{1}{\sqrt{2\pi}} \int_z^{\infty} e^{-\frac{t^2}{2}} \cdot dt$ - which is the integral of a normally distributed individual plant responses. It is effectively the cumulative response of the population).

Ryegrass plants show a double response to microwave treatment (Figure 8.4 – top left), with some efficacy at low application energy; however, 100% mortality required 600 J $cm^{-2}$ of microwave energy at the soil surface. This is probably because grasses have their apical meristem at the base of the plant in or near the soil surface, where it is slightly protected from microwave heating by the leaves above and the surrounding soil. This is important to consider when designing an effective microwave applicator system for microwave weed management in cropping systems. The dose response for ryegrass is described by:

$$S = 0.58 \bullet erfc[0.013(E + 1.24 \times 10^{-7})] + 0.174 \bullet erfc[0.0097(E - 448.4)] \tag{8.2}$$

The $r^2$ value for this function is 0. 72. The results from the second repetition of this experiment were also the same as the first.

Figure 8.4 shows the response curves for four of the species tested during these experiments. It is apparent that some species are more susceptible to microwave treatment than others. Table 8.1 summarises the response parameters for the different test species and indicates the lethal doses for 50% ($LD_{50}$) and 90% ($LD_{90}$) weed control.

### 8.2.2 Moving Trailer Experiments

A novel antenna, which redirects the microwave energy horizontally rather than vertically, was developed (Brodie, Torkovnikov, & Farrell, 2016). This antenna was connected to the prototype trailer (Figure 8.3) so that the microwave applicators was

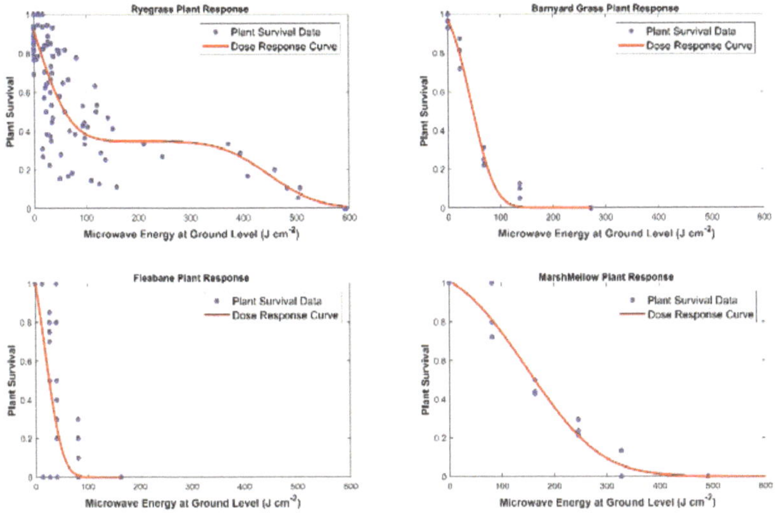


**Figure 8.4:** Dose response curves for microwave treatment of four species of weed plant using a horn antenna. Dose curves are calculated using equations (8.1) and (8.2).

**Table 8.1:** Equation coefficients, goodness of fit ($R^2$), $LD_{50}$, and $LD_{90}$ for weed plant survival as a function of microwave energy applied to the soil surface.

| Species | Coefficients | | | | | | $R^2$ | $LD_{50}$ (J cm⁻²) | $LD_{90}$ (J cm⁻²) |
|---|---|---|---|---|---|---|---|---|---|
| | a | b | c | d | e | f | | | |
| Annual Ryegrass | 0.576 | 0.013 | 1.24E-07 | 0.174 | 0.01 | 448.4 | 0.72 | 60 | 480 |
| Barnyard grass | 0.54 | 0.02 | 44.7 | — | — | — | 0.98 | 48 | 91 |
| Barley Grass | 0.56 | 0.022 | 40.92 | — | — | — | 0.99 | 45 | 84 |
| Brome Grass | 0.58 | 0.012 | 65.19 | — | — | — | 0.98 | 76 | 148 |
| Feathertop Rhodes Grass | 0.5 | 0.045 | 41.02 | — | — | — | 0.99 | 41 | 61 |
| Fleabane | 0.52 | 0.04 | 37.57 | — | — | — | 0.99 | 39 | 61 |
| Marshmallow | 0.55 | 0.0064 | 150.1 | — | — | — | 0.98 | 163 | 297 |
| Paddy Mellon | 0.52 | 0.047 | 33.92 | — | — | — | 0.99 | 35 | 53 |
| Wild Oats | 0.54 | 0.024 | 41.23 | — | — | — | 0.97 | 44 | 80 |
| Wild Radish | 0.52 | 0.017 | 64.53 | — | — | — | 0.99 | 67 | 118 |

about 15 – 20 mm above the ground. With this arrangement, the trailer was towed over short sections of Kikuyu grass (*Pennisetum clandestinum*).

The treated sections were about 1.5 to 2.0 m long, on flat surfaces, to avoid accidental damage to the wave-guides due to engagement with the soil surface. The travel time for these strips was between 7 and 10 seconds, equating to about 720 m $h^{-1}$. The treated strips were evaluated about 4 days after treatment. Only one of the four microwave generators were used during these experiments.

Thermal images revealed that the grass achieved a maximum temperature of 61°C (Figure 8.5). There was audible crackling of the grass as the applicator moved along the strip, indicating that micro-steam explosions were occurring in the grass stems due to rapid microwave heating (Brodie, 2007; Brodie et al., 2011). The grass was wet and wilted immediately after treatment (Figure 8.6), and there was evidence of scorching later, on the same day that the treatment was applied. After four days, the treated strips were quite evident in the grassed area (Figure 8.7). There was 100% control of the grass along all treated strips, with the treated strip being 100 mm wide.

Based on using a 2 kW microwave generator to treat a strip 720 m long by 0.1 m wide in one hour, the energy density for this treatment is equivalent to 10 J $cm^{-2}$ for 100% control of Kikuyu grass.

Other experiments, involving ryegrass and general weedy areas with multiple species have been conducted with the new antenna, with the same results.

The purpose of this experiment was to demonstrate mobility of the system. It is apparent from the results that movement of the applicator over the ground is possible and treatment is effective. The optimal operating speed of the system over grass covered ground is yet to be determined; however, these results are very encouraging. The optimal travel speed is yet to be determined; however, there was 100% plant kill in all the test strips at this speed. It is also important to note that the treatment strips are very clearly defined in the grass; therefore, with the correct guiding technology, microwave treatment can be very selective and has no effect adjacent plants. This should lend itself to effective inter-row weed control, without affecting the adjacent crop.

## 8.3 Conclusion

Microwave treatment can kill weed plants, above the soil surface. Static experiments indicate that the required energy varies considerably from species to species. The average lethal dose needed to 90% efficacy across all the tested species (Tab. 8.1) is 137 J $cm^{-2}$; however, the early experiments with the novel antenna indicate that 10 J $cm^{-2}$ is sufficient to kill emerged plants using this system. This equates to 1.0 GJ $ha^{-1}$. This is comparable to the total energy associated with herbicide weed management (see Chapter 4). Allowing an efficiency of between 75 and 90% for conversion of electrical energy into microwave energy, this equates to a treatment energy of between 1.1 and 1.3 GJ $ha^{-1}$.

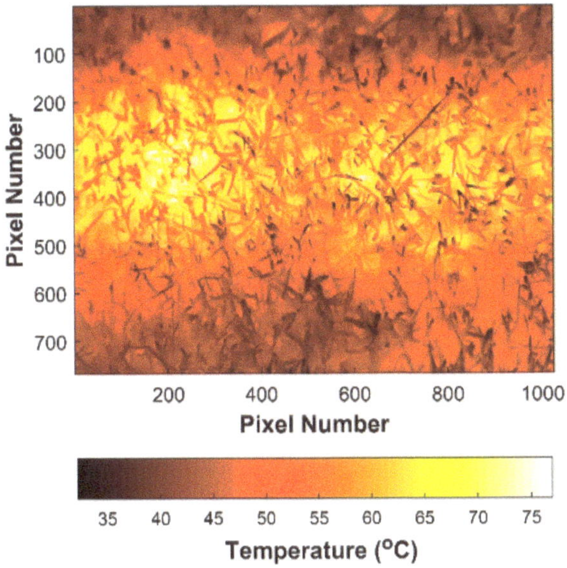


**Figure 8.5:** Thermal image of treated strip (Treatment = 1.5 m in 7 seconds), immediately after treatment.

**Figure 8.6:** Visible light image of the treated strip, shown in Figure 8.5.

**Figure 8.7:** Image of four treated strips of kikuyu grass, taken (left) 4 days after treatment and (right) 20 days after treatment.

## 8.4 References

Ark, P. A., & Parry, W. (1940). Application of High-Frequency Electrostatic Fields in Agriculture. The Quarterly Review of Biology, 15(2), 172-191.

Bebawi, F. F., Cooper, A. P., Brodie, G. I., Madigan, B. A., Vitelli, J. S., Worsley, K. J., & Davis, K. M. (2007). Effect of microwave radiation on seed mortality of rubber vine (Cryptostegia grandiflora R.Br.), parthenium (Parthenium hysterophorous L.) and bellyache bush (Jatropha gossypiifolia L.). Plant Protection Quarterly, 22(4), 136-142.

Brodie, G. (2007). Microwave treatment accelerates solar timber drying. Transactions of the American Society of Agricultural and Biological Engineers, 50(2), 389-396.

Brodie, G., Botta, C., & Woodworth, J. (2007). Preliminary investigation into microwave soil pasteurization using wheat as a test species. Plant Protection Quarterly, 22(2), 72-75.

Brodie, G., Hamilton, S., & Woodworth, J. (2007). An assessment of microwave soil pasteurization for killing seeds and weeds. Plant Protection Quarterly, 22(4), 143-149.

Brodie, G., Harris, G., Pasma, L., Travers, A., Leyson, D., Lancaster, C., & Woodworth, J. (2009). Microwave soil heating for controlling ryegrass seed germination. Transactions of the American Society of Agricultural and Biological Engineers, 52(1), 295-302.

Brodie, G., & Hollins, E. (2015). The Effect of Microwave Treatment on Ryegrass and Wild Radish Plants and Seeds. Global Journal of Agricultural Innovation, Research & Development, 2(1), 16-24.

Brodie, G., Jacob, M. V., Sheehan, M., Yin, L., Cushion, M., & Harris, G. (2011). Microwave modification of sugar cane to enhance juice extraction during milling. Journal of Microwave Power and Electromagnetic Energy, 45(4), 178-187.

Brodie, G., Ryan, C., & Lancaster, C. (2012). The effect of microwave radiation on Paddy Melon (Cucumis myriocarpus). International Journal of Agronomy, 2012, 1-10.

Brodie, G., Torkovnikov, G., & Farrell, P. (2016). Australia Patent No. 2016905272. A. P. Office.

Diprose, M. F., Benson, F. A., & Willis, A. J. (1984). The Effect of Externally Applied Electrostatic Fields, Microwave Radiation and Electric Currents on Plants and Other Organisms, with Special Reference to Weed Control. Botanical Review, 50(2), 171-223.

Grundler, W., Keilmann, F., & Fröhlich, H. (1977). Resonant growth rate response of yeast cells irradiated by weak microwaves. Physics Letters A, 62(6), 463-466. doi:https://doi.org/10.1016/0375-9601(77)90696-X

Hightower, N. C., Burdette, E. C., & Burns, C. P. (1974). Investigation of the use of microwave energy for weed seed and wood products insect control, Final Technical report. Project E-230-901. Retrieved from

Horikoshi, S., Hasegawa, Y., & Suzuki, N. (2017). Benefitting of Plants Using Microwave Genetic Activation Method and its Applications. Paper presented at the IMPI's 51st Annual Microwave Power Symposium (IMPI 51), Miami, FL, USA.

Magone, I. (1996). The effect of electromagnetic radiation from the Skrunda radio location station on Spirodela polyrhiza (L.) Schleiden cultures. Science of theTotal Environment, 180(1), 75–80.

Petrov, I. Y., Moiseeva, T. V., & Morozova, E. V. (1991). A possibility of correction of vital processes in plant cell with microwave radiation. Paper presented at the 1991 IEEE International Symposium on Electromagnetic Compatibility, Cherry Hill, NJ, USA.

Skiles, J. W. (2006). Plant response to microwaves at 2.45 GHz. Acta Astronautica, 58(5), 258-263. doi:DOI: 10.1016/j.actaastro.2005.12.007

Wayland, J., Merkle, M., Davis, F., Menges, R. M., & Robinson, R. (1975). Control of weeds with UHF electromagnetic fields. Unkrautbekämpfung mit elektromagnetischen UHF-Feldern., 15(1), 1-5.

Zapata-Sierra, A. J., & Manzano-Agugliaro, F. (2017). Controlled deficit irrigation for orange trees in Mediterranean countries. Journal of Cleaner Production, 162(Supplement C), 130-140. doi:https://doi.org/10.1016/j.jclepro.2017.05.208

# 9 The Potential of Microwave Soil Treatment to Kill Weed Seeds

## 9.1 Introduction

Weed management is closely associated with weed seed bank management. No-till systems have the largest portion (90%) of their weed seed bank in the top 0 to 5 cm of soil (Swanton, Shrestha, Knezevic, Roy, & Ball-Coelho, 2000). Peltzer and Matson (2002) studied the seed bank longevity of five annual weed species: annual ryegrass (*Lolium rigidum*), barley grass (*Hordeum leporinum [H. murinum subsp. leporinum]*), wild radish (*Raphanus raphanistrum*), wild oat (*Avena fatua*) and wall fumitory (*Fumaria muralis*) in the southwest land division of Western Australia. They ensured that no seed set was allowed in their trial plots. Barley grass did not persist in the soil while the seeds of the other species persisted in the soil for several years. Soil fumigation is sometimes used to quickly deplete the soil seed bank before planting high value crops (Gallandt, Fuerst, & Kennedy, 2004).

Soil fumigants, such as Methyl bromide (bromomethane), have been used widely in agriculture since the 1940's. They can eradicate nematodes, plant pathogens, weed seeds and insects in the soil, largely due to: their wide spectrum of activity against soil biota; their ability to penetrate the fumigated zones; and their ease of application (Ibekwe, Papiernik, & Yang, 2010). Soil fumigants are used for many commercial crops, including: strawberries; tomatoes; peppers; eggplants; tobacco; ornamentals; nursery stocks; vines; and turves (Ibekwe et al., 2010). Soil fumigants are hazardous to work with and Methyl bromide is being phased out in most countries (Sydorovych et al., 2006). This has prompted a search for alternative methods of soil fumigation for controlling weeds, insects, nematodes, and other plant pathogens.

Davis (1973) showed that microwave treatment kills dormant seeds. Seed volume significantly correlates with susceptibility to microwave damage; however, this may not be due to direct interaction of the seeds with microwave energy, but rather it may be due to better heat transfer from surrounding soil (Nelson, 1996) because the radar cross section (Wolf, Vaughn, Harris, & Loper, 1993) of seeds is very small and therefore direct absorption of microwave energy will also be small. Experiments have demonstrated that microwaves interact with soil rather than directly with the seeds; however, heat is transferred from the soil to the seeds (G. Brodie, Botta, & Woodworth, 2007). This chapter explores the potential for microwave soil treatment to control weed seeds in soil.

## 9.2 Seed Treatment

Brodie (Brodie, Botta, et al., 2007; Brodie, Hamilton, & Woodworth, 2007; Brodie et al., 2009; Brodie & Hollins, 2015; Brodie, Pasma, Bennett, Harris, & Woodworth, 2007) has conducted several weed seed experiments, were either air dry or moist soil (20% moisture by volume) was layered into pots with sets of between 10 to 25 seeds placed into paper envelopes at depths of 0, 2, 5, 10 and 20 cm within each pot. Pots were treated for 0, 2, 5, 10, 30, 60, or 120 seconds using a prototype microwave system fed from a conventional 600 W microwave oven magnetron into a pyramidal horn antenna. The paper envelopes allowed easy seed recovery from the soil after treatment so that seeds could be germinated in a growth cabinet for viability assessment. The following species were evaluated: annual ryegrass (*Lolium rigidum* L.); perennial ryegrass; bellyache bush (*Jatropha gossypiifolia* L.); giant sensitive tree, catclaw plant or bashful plant (*Mimosa pigra* L.); parthenium (*Parthenium hysterophorous* L.); rubber vine (*Cryptostegia grandiflora* R.Br.); wild radish (*Raphanus raphanistrum* L.), and wild oats (*Avena fatua* L.).

The first prototype microwave system, described in Chapter 8, was used to treat seeds in fine builder's sand. Dry sand has the lowest dielectric constant at microwave frequencies of any air-dry soil type (Velazquez-Marti, 2005; Von Hippel, 1954); therefore, sand was chosen to represent the "worst case scenario" for microwave treatment of soil seed banks. The sand was sieved through a 1 mm soil sieve to ensure homogeneity of soil behaviour during the experiments. The sand was air-dried during summer and then stored in a warm dry environment until used in the experiments. The experiment also considered the effect of soil moisture by having two moisture levels: air dry and 20% water by volume.

## 9.3 Modelling Seed Response

The relationships between applied microwave energy and seed survival were fitted to a dose response surface of the form:

$$S=a\bullet \mathrm{erfc}[b\bullet(\Psi\bullet e^{-2cd}-f)] \tag{9.1}$$

Where d is the depth of the seeds in the soil profile (m), and a, b, c (field attenuation rate in soil) and f (median seed response) are constants to be experimentally determined for each species. As before, the Gaussian error function assumes a normally distributed seed response to microwave energy. Including a term with the form $E\bullet e^{-\alpha D}$ will account for the natural attenuation of the microwave energy with depth in the soil.

Some examples of the fitting of these curves to measured data are shown in Figure 9.1. In all cases the microwave energy dose is based on the energy density at ground level. Table 9.1 summarises the dose responses of several species

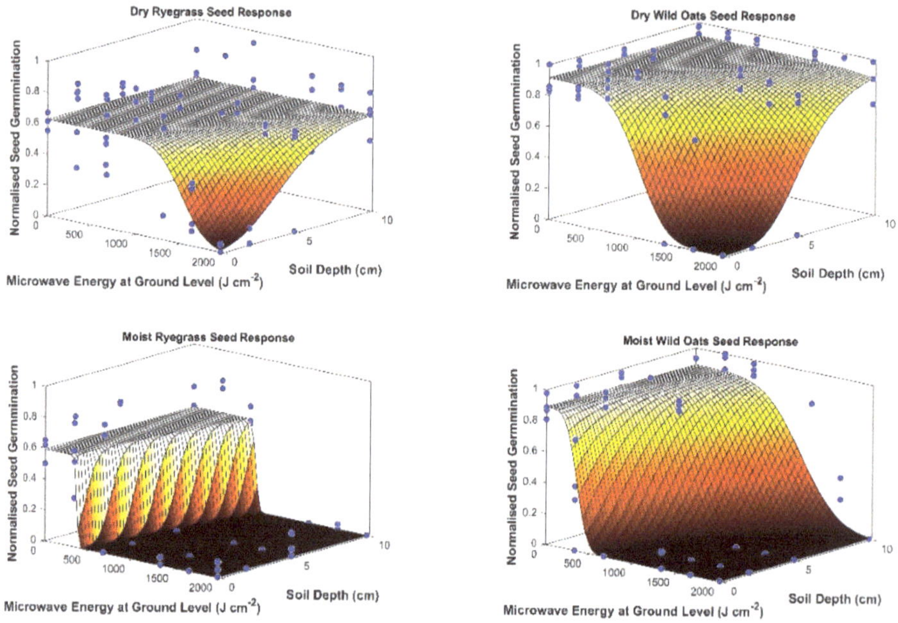


**Figure 9.1:** Dose responses of ryegrass and wild oats seeds as a function of soil moisture, microwave energy at ground level and burial depth in soil.

subjected to experimental investigation by Brodie et al. (Brodie, Botta, et al., 2007; Brodie, Hamilton, et al., 2007; Brodie et al., 2009; Brodie & Hollins, 2015; Brodie, Pasma, et al., 2007).

Applying equation (9.1) to the work by Menges and Wayland (1974), reported in Chapter 6, reveals that the $LD_{50}$ and $LD_{90}$ needed to kill 50% and 90% of the seeds at 2 cm burial depth in irrigated soil was 78.5 and 98.4 J $cm^{-2}$ respectively. The $LD_{50}$ and $LD_{90}$ for their non-irrigated experiments were 94.2 and 136.8 J $cm^{-2}$ respectively. These values are much lower than in sand (Tab. 9.1), but are like the values for Wild Radish, because this experiment was conducted in moist soil rather than sand. Sand heats more slowly than soil and particularly clay, as was pointed out in Chapter 7.

## 9.4 Treatment Efficiency

Using equations derived for the temperature distribution in microwave heated soil (Brodie, 2008; Brodie, Hamilton, et al., 2007), the performance index, which was developed by Gay, Piccarolo, Ricauda Aimonino, and Tortia (2010) to explore the

**Table 9.1:** Equation coefficients, and goodness of fit ($R^2$) for weed seed survival as a function of microwave energy applied to the sand surface, seed burial depth, and soil moisture status (Source: G. Brodie, Hamilton, et al., 2007; G. Brodie et al., 2009; Graham Brodie & Hollins, 2015; Menges & Wayland, 1974).

| Species | Sand Moisture | | | | | | | | | | | | | |
|---|---|---|---|---|---|---|---|---|---|---|---|---|---|---|
| | Dry | | | | | $LD_{50}$ * | $LD_{90}$ * | Wet | | | | | $LD_{50}$ * | $LD_{90}$ * |
| | a | b | c | f | $R^2$ | (J cm$^{-2}$) | (J cm$^{-2}$) | a | b | c | f | $R^2$ | (J cm$^{-2}$) | (J cm$^{-2}$) |
| Annual Ryegrass | 0.31 | 0.0033 | 0.06 | 1521 | 0.52 | 1339 | 1737 | 0.30 | 0.0456 | 0.07 | 355.40 | 0.91 | 341 | 371 |
| Bellyache bush | 0.50 | 0.0020 | 0.08 | 664.5 | 0.90 | 666 | 1117 | 0.56 | 0.0033 | 0.15 | 255.8 | 0.87 | 287 | 546 |
| Parthenium | 0.55 | 0.0011 | 0.22 | 762.8 | 0.79 | 842 | 1614 | 0.97 | 0.0012 | 0.08 | 0.0001 | 0.73 | 375 | 940 |
| Perrenial Ryegrass | 0.41 | 0.0027 | 0.06 | 1400 | 0.78 | 1330 | 1708 | 0.43 | 0.0148 | 0.13 | 240.5 | 0.94 | 232 | 299 |
| Rubber vine | 0.49 | 0.0020 | 0.03 | 936.6 | 0.58 | 1386 | 310 | 0.61 | 0.0015 | 0.03 | 406.00 | 0.86 | 513 | 1046 |
| Wild Oats | 0.46 | 0.0028 | 0.12 | 1006 | 0.76 | 981 | 1326 | 0.45 | 0.0074 | 0.12 | 346.8 | 0.84 | 334 | 465 |
| Wild Radish ¥ | - | - | - | - | - | - | - | 0.16 | 0.1083 | 0.12 | 74.25 | 0.72 | 65 | 78 |
| Menges and Wayland (1974) ¥ | 0.3392 | 0.02805 | 0.104 | 109.8 | 0.63 | 94 | 137 | 0.349 | 0.05884 | 0.2834 | 84.51 | 0.69 | 79 | 99 |

* The $LD_{50}$ and $LD_{90}$ values represent the soil surface energy doses needed to kill 50% and 90% of the seeds at 2 cm burial depth, respectively.
¥ These experiments were conducted in moist clay soil rather than sand, like the other experiments

efficiency of steam treatment, can be derived. The performance of microwave soil treatment using a horn antenna to irradiate the soil is given by:

$$I = \frac{n\omega\varepsilon_O\kappa'' E_O^2}{64 t_f k \gamma \alpha^4}\left(e^{4\gamma\alpha^2 t_f} - 1\right) \tag{9.2}$$

Where n is an amplitude scaling factor for simultaneous heat and moisture movement during microwave heating (Brodie, 2007; Henry, 1948), ω is the angular velocity of the microwave field (Rad $s^{-1}$), $\varepsilon_0$ is the permittivity of free space, κ" is the dielectric loss factor of the soil, $E_0$ is the electric field strength at the soil surface (V $m^{-1}$), $t_f$ is the microwave heating time (s), k is the thermal conductivity of the soil (W $m^{-1}$ $K^{-1}$), γ is the combined heat and moisture diffusivity coefficient for the soil, and α is the microwave attenuation factor in the soil ($m^{-1}$).

Equation (9.2) can be used to estimate the heating efficiency index for clay soil, which is being heated for 120 seconds using a 200 W, 2.45 GHz microwave source feeding into a horn antenna with aperture dimensions 110 mm by 55 mm. The heating efficiency index is 80, which is almost twice that of steam treatment, as estimated by Gay et al. (2010). This improvement in the heating efficiency index is probably linked to the rapidity of microwave heating, compared to steam treatment, because microwave heating is a spatial energy transfer phenomenon while steam heating depends on convective heat transfer.

## 9.5 Conclusion

Several experiments explored the interaction between microwave energy and seed depth in the soil. Seeds in the top layer of soil were affected by microwave treatment; however, this effect diminished with burial depth. Dry soil requires more energy to treat than irrigated soil; however, effective treatment is deeper in dry soil compared with dry soil. The $LD_{90}$ for effective seed treatment to 2 cm depth varies between about 78 and 100 J $cm^{-2}$. This equates to 10 GJ $ha^{-1}$. Soil treatment for weed seed control therefore requires about ten times the energy than killing emerged weeds (see Chapter 8); however, it may have application in some high value horticultural crops, which already use soil fumigation prior to planting the commercial crop.

## 9.6 References


Brodie, G. (2007). Simultaneous heat and moisture diffusion during microwave heating of moist wood. Applied Engineering in Agriculture, 23(2), 179-187.

Brodie, G. (2008). The influence of load geometry on temperature distribution during microwave heating. Transactions of the American Society of Agricultural and Biological Engineers, 51(4), 1401-1413.

Brodie, G., Botta, C., & Woodworth, J. (2007). Preliminary investigation into microwave soil pasteurization using wheat as a test species. Plant Protection Quarterly, 22(2), 72-75.

Brodie, G., Hamilton, S., & Woodworth, J. (2007). An assessment of microwave soil pasteurization for killing seeds and weeds. Plant Protection Quarterly, 22(4), 143-149.

Brodie, G., Harris, G., Pasma, L., Travers, A., Leyson, D., Lancaster, C., & Woodworth, J. (2009). Microwave soil heating for controlling ryegrass seed germination. Transactions of the American Society of Agricultural and Biological Engineers, 52(1), 295-302.

Brodie, G., & Hollins, E. (2015). The Effect of Microwave Treatment on Ryegrass and Wild Radish Plants and Seeds. Global Journal of Agricultural Innovation, Research & Development, 2(1), 16-24.

Brodie, G., Pasma, L., Bennett, H., Harris, G., & Woodworth, J. (2007). Evaluation of microwave soil pasteurization for controlling germination of perennial ryegrass (Lolium perenne) seeds. Plant Protection Quarterly, 22(4), 150-154.

Davis, F. S., Wayland, J. R. and Merkle, M. G. (1973). Phytotoxicity of a UHF Electromagnetic Field. Nature, 241(5387), 291-292.

Gallandt, E. R., Fuerst, E. P., & Kennedy, A. C. (2004). Effect of tillage, fungicide seed treatment, and soil fumigation on seed bank dynamics of wild oat (Avena fatua). Weed Science, 52(4), 597-604.

Gay, P., Piccarolo, P., Ricauda Aimonino, D., & Tortia, C. (2010). A high efficiency steam soil disinfestation system, part I: Physical background and steam supply optimisation. Biosystems Engineering, 107(2), 74-85. doi:http://dx.doi.org/10.1016/j.biosystemseng.2010.07.003

Henry, P. S. H. (1948). The diffusion of moisture and heat through textiles. Discussions of the Faraday Society, 3, 243-257.

Ibekwe, A. M., Papiernik, S. K., & Yang, C. H. (2010). Influence of soil fumigation by methyl bromide and methyl iodide on rhizosphere and phyllosphere microbial community structure. Journal of Environmental Science & Health, Part B -- Pesticides, Food Contaminants, & Agricultural Wastes, 45(5), 427-436. doi:10.1080/03601231003800131

Menges, R. M., & Wayland, J. R. (1974). UHF electromagnetic energy for weed control in vegetables. Weed Science, 22(6), 584-590.

Nelson, S. O. (1996). A review and assessment of microwave energy for soil treatment to control pests. Transactions of the ASAE, 39(1), 281-289.

Peltzer, S. C., & Matson, P. T. (2002). How fast do the seedbanks of five annual cropping weeds deplete in the absence of weed seed input? Paper presented at the 13th Australian Weeds Conference: weeds "threats now and forever?", Perth, Western Australia.

Swanton, C. J., Shrestha, A., Knezevic, S. Z., Roy, R. C., & Ball-Coelho, B. R. (2000). Influence of tillage type on vertical weed seedbank distribution in a sandy soil. Canadian Journal of Plant Science, 80(2), 455-457. doi:10.4141/P99-020

Sydorovych, O., Safley, C. D., Ferguson, L. M., Poling, E. B., Fernandez, G. E., Brannen, P. A., et al. (2006). Economic evaluation of methyl bromide alternatives for the production of strawberries in the southeastern United States. HortTechnology, 16(1), 118 -128.

Velazquez-Marti, B., Gracia-Lopez, C., and Plaza-Gonzalez, P. J. (2005). Determination of dielectric properties of agricultural soil. Biosystems Engineering, 91(1), 119-125.

Von Hippel, A. R. (1954). Dielectric Materials and Applications. Cambridge: M.I.T. Press.

Wolf, W. W., Vaughn, C. R., Harris, R., & Loper, G. M. (1993). Insect radar cross-section for aerial density measurement and target classification. Transactions of the American Society of Agricultural and Biological Engineers, 36(3), 949-954.

## 9.7 Appendix - Analysis of Energy in Soil after Microwave Treatment

$$dE = \rho CT \cdot dV \tag{9.3}$$

Where r is the density of the soil (kg m-3), C is the thermal capacity of the soil (J kg$^{-1}$ K$^{-1}$), T is temperature (K) and dV is a small volume.

Based on earlier modelling of microwave heating (Brodie, 2008), the temperature distribution in the soil as a result of microwave heating with a horn antenna in contact with the soil surface is given by:

$$T = \frac{n\omega\varepsilon_o\kappa''E_o^2}{16k\alpha^2}\left(e^{4\gamma\alpha^2 t}-1\right)\left[e^{-2\alpha z}+\left(\frac{h}{k}+2\alpha\right)\cdot z\cdot e^{\frac{-z^2}{4\gamma t}}\right]\cdot Cos^2\left(\frac{\pi y}{A}\right) \cdot\left[erf\left(\frac{x+\frac{H}{2}}{2\sqrt{\gamma t}}\right)-erf\left(\frac{x-\frac{H}{2}}{2\sqrt{\gamma t}}\right)\right] \tag{9.4}$$

Therefore, the total energy in the microwave heated soil is:

$$E = \int_V \rho CT \cdot dV = \rho C \int_V T \cdot dV \tag{9.5}$$

This can be developed using a piece wise approach.

Firstly:

$$\begin{aligned}\int_0^\infty \left[e^{-2\alpha z}+\left(\frac{h}{k}+2\alpha\right)\cdot z\cdot e^{\frac{-z^2}{4\gamma t}}\right]\cdot dz \\ = \left[\frac{e^{-2\alpha z}}{-2\alpha}-2\gamma t\left(\frac{h}{k}+2\alpha\right)\cdot e^{\frac{-z^2}{4\gamma t}}\right]_{z=0}^{\infty} \\ = \frac{1}{2\alpha}+2\gamma t\left(\frac{h}{k}+2\alpha\right)\end{aligned} \tag{9.6}$$

Secondly:

$$\int_{-A/2}^{A/2} Cos^2\left(\frac{\pi y}{A}\right) \cdot dy \tag{9.7}$$

Let $u = \frac{\pi y}{A}$. When $y = \frac{A}{2}$, $u = \frac{\pi}{2}$ and $dy = \frac{A}{\pi} \cdot du$

Therefore:

$$\begin{aligned} \int_{-A/2}^{A/2} Cos^2\left(\frac{\pi y}{A}\right) \cdot dy &= \frac{A}{\pi}\int_{-\pi/2}^{\pi/2} Cos^2(u) \cdot du \\ &\frac{A}{2\pi}\left[u + Sin(u) \cdot Cos(u)\right]_{-\pi/2}^{\pi/2} \\ &= \frac{A}{2} \end{aligned} \tag{9.8}$$

Thirdly:

$$\begin{aligned} \int_{-\infty}^{\infty}\left[erf\left(\frac{x+\frac{H}{2}}{2\sqrt{\gamma t}}\right) - erf\left(\frac{x-\frac{H}{2}}{2\sqrt{\gamma t}}\right)\right] \cdot dx &= 2\int_0^{\infty} 1 - erf\left(\frac{x-\frac{H}{2}}{2\sqrt{\gamma t}}\right) \cdot dx \\ &= 2\int_0^{\infty} erfc\left(\frac{x-\frac{H}{2}}{2\sqrt{\gamma t}}\right) \cdot dx \end{aligned} \tag{9.9}$$

Let $u = \frac{x-\frac{H}{2}}{2\sqrt{\gamma t}}$. When $x = 0; u = \frac{-H}{4\sqrt{\gamma t}}$. When $x = \infty; u = \infty$. $dx = 2\sqrt{\gamma t} \cdot du$

Therefore:

$$\begin{aligned} 2\int_0^{\infty} erfc\left(\frac{x-\frac{H}{2}}{2\sqrt{\gamma t}}\right) \cdot dx &= 4\sqrt{\gamma t}\int_{\frac{-H}{4\sqrt{\gamma t}}}^{\infty} erfc(u) \cdot du \\ &= 4\sqrt{\gamma t}\left[u \cdot erfc(u) - \frac{e^{-u^2}}{\sqrt{\pi}}\right]_{\frac{-H}{4\sqrt{\gamma t}}}^{\infty} \end{aligned} \tag{9.10}$$

$$= 4\sqrt{\gamma t}\left[0 + 0 - \frac{-H}{4\sqrt{\gamma t}} \cdot erfc\left(\frac{-H}{4\sqrt{\gamma t}}\right) + \frac{e^{\frac{-H^2}{16\gamma t}}}{\sqrt{\pi}}\right] = 4\sqrt{\gamma t}\left[\frac{II}{4\sqrt{\gamma t}} \cdot erfc\left(\frac{-H}{4\sqrt{\gamma t}}\right) + \frac{e^{\frac{-H^2}{16\gamma t}}}{\sqrt{\pi}}\right]$$

Note that γ is very small, therefore: $\frac{e^{\frac{-H^2}{16\gamma t}}}{\sqrt{\pi}}$ is practically zero.
Therefore:

$$\int_{-\infty}^{\infty}\left[erf\left(\frac{x+\frac{H}{2}}{2\sqrt{\gamma t}}\right)-erf\left(\frac{x-\frac{H}{2}}{2\sqrt{\gamma t}}\right)\right]\cdot dx = H\cdot erfc\left(\frac{-H}{4\sqrt{\gamma t}}\right) \tag{9.11}$$

Combining these elements together gives:

$$E = \rho C\frac{n\omega\varepsilon_o\kappa''E_o^2}{16k\alpha^2}\left(e^{4\gamma\alpha^2 t}-1\right)\left[\frac{1}{2\alpha}+2\gamma t\left(\frac{h}{k}+2\alpha\right)\right]\cdot\frac{A}{2}\cdot H\cdot erfc\left(\frac{-H}{4\sqrt{\gamma t}}\right)$$

Or

$$E = \rho C\frac{n\omega\varepsilon_o\kappa''E_o^2}{16k\alpha^2}\left(e^{4\gamma\alpha^2 t}-1\right)\left[\frac{1}{2\alpha}+2\gamma t\left(\frac{h}{k}+2\alpha\right)\right]\cdot\frac{A}{2}\cdot H\cdot\left[2-erfc\left(\frac{H}{4\sqrt{\gamma t}}\right)\right]$$

Because γ is very small this equation simplifies to:

$$E = \rho C\frac{n\omega\varepsilon_o\kappa''E_o^2}{16k\alpha^2}\left(e^{4\gamma\alpha^2 t}-1\right)\frac{1}{2\alpha}\cdot\frac{A}{2}\cdot H\cdot 2$$

Or

$$E = \rho C\frac{n\omega\varepsilon_o\kappa''E_o^2}{32k\alpha^3}\left(e^{4\gamma\alpha^2 t}-1\right)\cdot A\cdot H \tag{9.12}$$

Gay et al. (2010) developed a scalar index to measure efficiency for steam heating

$$I = \frac{1}{V\cdot t_f}\int_0^{t_f}\int_V T\cdot dV\cdot dt \tag{9.13}$$

Or

$$I = \frac{1}{Vt_f}\int_0^{t_f}\frac{n\omega\varepsilon_o\kappa''E_o^2}{32k\alpha^3}\left(e^{4\gamma\alpha^2 t}-1\right)\cdot A\cdot H\cdot dt$$

This becomes:

$$I = \frac{n\omega\varepsilon_o\kappa''E_o^2}{32Vt_fk\alpha^3}\cdot A\cdot H\int_0^{t_f}\left(e^{4\gamma\alpha^2 t}-1\right)\cdot dt \tag{9.14}$$

Evaluating the integral gives:

$$I = \frac{n\omega\varepsilon_o\kappa''E_o^2}{32Vt_fk\alpha^3}\cdot A\cdot H\left(\frac{e^{4\gamma\alpha^2t_f}}{4\gamma\alpha^2} - t_f - \frac{1}{4\gamma\alpha^2}\right)$$

Again, because γ is very small:

$$I = \frac{n\omega\varepsilon_o\kappa''E_o^2}{128Vt_fk\gamma\alpha^5}\cdot A\cdot H\left(e^{4\gamma\alpha^2t_f} - 1\right)$$

The volume of soil is: $V = \frac{AH}{2\alpha}$; therefore

$$I = \frac{n\omega\varepsilon_o\kappa''E_o^2}{64t_fk\gamma\alpha^4}\left(e^{4\gamma\alpha^2t_f} - 1\right) \tag{9.15}$$

# 10 The Effect of Microwave Treatment on Soil Biota

## 10.1 Introduction

The previous chapter demonstrated that microwave treatment of soil can kill weed seeds (Brodie, Hamilton, & Woodworth, 2007; Brodie & Hollins, 2015; F. Davis, 1975; F. S. Davis, 1974; F. S. Davis, Wayland, & Merkle, 1971; F. S. Davis, Wayland, J. R. and Merkle, M. G., 1973). Experiments have demonstrated that raising the soil temperature above 80°C will kill seeds of most of crops such as wheat (Brodie et al., 2007), ryegrass (Brodie et al., 2009), rubber vine, parthenium, bellyache bush (Bebawi et al., 2007), Prickly Paddy Melon (Brodie, Ryan, & Lancaster, 2012a), wild oats (Brodie, Ryan, & Lancaster, 2012b), white clover, and hemlock (Brodie, Harris, & Torgovnikov, 2014); however, one of the other potential applications of microwave soil treatment is as a substitute for soil fumigation.

Soil fumigation is used, not only to manage weed emergence, but to also manage other soil biota, such as bacteria, fungi and nematodes (Samtani et al., 2012). Soil biota consist of mainly living or residues of macro and micro-organisms (fungi, bacteria, algae, cyanobacteria, archaea, and nematodes, earthworms, protozoa, mites, insects) as well as plant residues. Soil biota is affected by interaction of abiotic and biotic factors. Soil heating in various forms affect soil biota. Previous studies especially with soil solarisation has shown some variable effect on soil biota; however, there are few reports which have investigated the effect of microwave treatment on soil biota.

Other studies have revealed that the amount of microwave energy required to kill emerged broad leaf weed plants is at least an order of magnitude less than the energy needed for seed inactivation in the top layers of soil (Brodie et al., 2012b). Microwave soil treatment also kills nematodes (Rahi & Rich, 2008, 2011). Studies of microwave effects on soil invertebrates are rare. It has been demonstrated that long exposure of earth worms (*Eisenia fetida*), in the absence of soil, to low intensity microwave fields (23 V $m^{-1}$ for 2 hours at frequencies of 900 MHz and 1.8 GHz) induced measurable DNA damage; however, lower intensity fields (10 V $m^{-1}$ for 2 hours at frequencies of 900 MHz and 1.8 GHz) had no measurable effect on their DNA (Malarić, Štambuk, Šrut, & Tkalec, 2008). This study was focused on the effect of long term exposure to low power electronic communication systems rather than microwave soil treatment. It is anticipated that exposure of these organisms to intense microwave fields will induce rapid internal heating, leading to death.

Speir et al. (1986) demonstrated that fungi are more susceptible to microwave soil treatment than bacteria. This has been verified by other researchers (Cooper & Brodie, 2009; Vela, Wu, & Smith, 1976; Wainwright, Killham, & Diprose, 1980). Microwave induced "heat shock" activation of bacterial and fungal spores has also been observed (Vela et al., 1976). Vela et al. (1976) also demonstrated that soil bacteria, bacterial spores, actinobacteria, fungi, nitrogen-fixing bacteria, and nitrifying bacteria were all resistant to over 40,000 J $cm^{-2}$ of microwave energy applied to the soil surface; no further reports are published in this area of research. Therefore, to understand the effect of microwave energy on soil biota this study was conducted.

## 10.2 Effect of Microwave Treatment of Soil Bacteria

In an experiment described in Cooper and Brodie (2009) predominantly clay soil was carefully layered into wooden boxes with pre-drilled soil sampling points at various locations down the side of the box (0, 2.5, 5, 10, 20 and 40 cm depths). These boxes were treated with five levels of microwave treatment (0, 2, 4, 8, and 16 minutes) using the first microwave oven prototype, described in Chapter 8. The prototype system was fitted with a 180 mm by 90 mm horn antenna mounted 50 mm above the soil surface. Soil samples were harvested from the six different depths in the soil profile after the soil had cooled to ambient temperature and bacteria were assessed using the pour plate method, which requires the use of 1, 0.1, 0.01, or 0.001 mL samples (Devine et al., 2007).

Only a very small portion of the total number of bacteria species present in soil can be cultured using the pour plate method and it is not clear which species may have survived to create colonies in the agar; however several outcomes were evident: there was considerable variability in bacterial numbers in the untreated soil; microwave treatment significantly reduces bacterial numbers, but did not 'sterilise' the soil; and the surface soil was most affected by microwave treatment, with the efficacy of microwave treatment diminishing with soil depth.

To better understand the impact of microwave treatment on bacteria, an experiment was conducted on a known species - *Escherichia coli* (*E. coli*). In an experiment described by Brodie et al. (2015), samples from a paddock at the Dookie campus of the University of Melbourne that was predominantly "Currawa Loam" were treated in an autoclave at 121°C at 15 psi for 20 minutes to sterilise the soil. Previously cultured *Escherichia coli* (*E. coli*) bacteria were inoculated into sterilised soil sub-samples.

One gram samples of the inoculated soil were placed inside small paper envelopes. Sterilised soil was used to fill twelve pots to a depth of 20 cm. Envelopes of inoculated soil were placed at various depths in the sterilised soil (suggested: 0, 2.5, 5, 10, and 20 cm). The trailer based prototype system, described in Chapter 8, was used to treat the pots at a range of 100 mm from the horn antenna, for treatment (treatment times: 0, 10, 30, and 120 seconds). The pour plate method was used to evaluate the impact of microwave treatment on bacterial populations.

Extended microwave treatment caused a $10^{-5}$ reduction in *E. coli* numbers in the top layer of soil; however, populations at greater depth were not significantly affected by microwave treatment.

The Normalised dose response curve, shown in Figure 10.1, has the form:

$$S=0.58 \bullet erfc(0.01 (\Psi \bullet e^{-0.34D} - 1.1\times10^{-6})) \qquad (10.1)$$

This is similar in form to the response surface for seed treatment, described in Chapter 9; however, the goodness of fit is only moderate ($R^2 = 0.47$). There was considerable variability in population densities in the control samples; however, the response to microwave treatment was clear. The $LD_{50}$ and $LD_{90}$ for the *E. coli* at 2 cm depth were 12.4 and 100.8 J $cm^{-2}$, respectively.

To further assess the impact of microwave treatment on soil bacterial numbers, soil profile samples were sampled randomly from a paddock at Dookie Campus of the University of Melbourne dominated by the Caniambo Loam soil type. A larger than needed volume of soil was removed carefully from the ground using a shovel so that the soil profile in the sample experienced minimal disturbance. Samples were then cut to fit into a 150 mm diameter pot using a knife and the soil was carefully placed into the pot to maintain the existing soil profile. If the profile was disturbed in this process, samples were discarded. The pots were placed into the Dookie campus glass house and watered.

The pots were subjected to four treatments (0, 30, 60, 120 seconds) using the trailer prototype system with the soil surface at a range of 100 mm from the horn antenna's aperture. Access points were made in the sides of the pots with a scalpel at 5 cm below the soil surface and at 10 cm below the soil surface. Soil samples were removed from the pots at these locations using an apple corer. Active bacteria assessments for the soil samples were carried out on $10^{-1}$ dilutions by applying fluorescent dye to a known volume of sample, mounting the sample in agar and viewing under an appropriate wavelength of light to facilitate fluorescence in the cells. Fluorescence microscopy is a rapidly expanding technique that is used in both medical and biological sciences. The technique has made it possible to identify cells and cellular components with a high degree of specificity (Cornea & Conn, 2014). Fluorescence can be induced in cells by addition of various chemicals: fluorescein diacetate for living cells and fluorescein isothiocyanate for non-living cells (Cornea & Conn, 2014). With adequate training, this technique can be used to determine the portions of living and dead specimens of bacteria extracted from the soil.

Analyses of the soil biota data revealed that microwave treatment significantly reduced the number of soil bacteria (Tab. 10.1) but did not completely sterilise the soil; however, bacterial numbers significantly increased after a month (Tab. 10.2) and ended significantly higher than at the start of the experiment.

Bacterial cells form the most concentrated C:N ratio of soil biota. Killing the cells through the microwave treatment provides extra nutrients for the remaining bacteria leading to an increase in the populations during the period following the treatment. Several of these experiments were repeated in different soils from the Dookie Campus of the university of Melbourne, with similar results. The combined response of bacteria

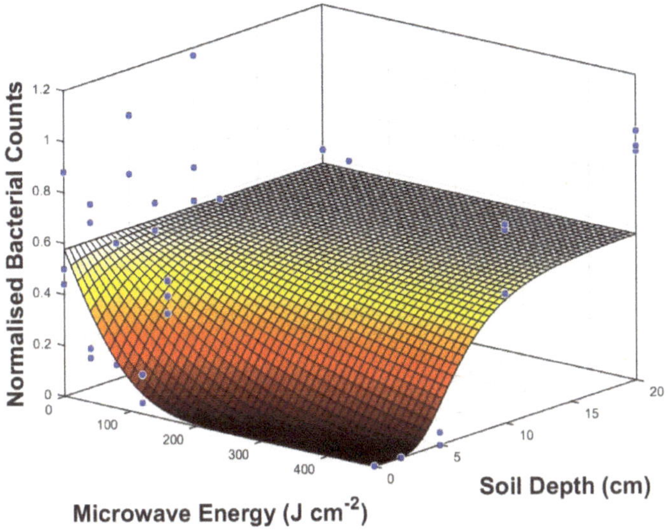


**Figure 10.1:** Normalised E. coli counts as a function of microwave energy and soil depth.

**Table 10.1:** Soil bacterial numbers shortly after microwave treatment (Entries in the table with different superscripts are significantly different to one another)(Source: Brodie et al., 2015).

| Soil Depth (cm) | Estimated Microwave Treatment (J cm$^{-2}$) | | | |
|---|---|---|---|---|
| | 0 | 150 | 300 | 600 |
| 0 | 6.20$^{a}$ | 5.57$^{a}$ | 4.73$^{ab}$ | 1.78$^{c}$ |
| 5 | 3.78$^{abc}$ | 4.71$^{ab}$ | 4.23$^{ab}$ | 1.18$^{c}$ |
| 10 | 4.06$^{ab}$ | 2.93$^{bc}$ | 3.87$^{abc}$ | 1.74$^{c}$ |
| LSD (P = 0.05) | | | | 2.60 |

from all these experiments is shown in Figure 10.2. Comparing Figure 10.1 and Figure 10.2 suggests that some species of bacteria are more resilient to microwave treatment than others, with the response surface of the combined bacterial response (Figure 10.2) being described by:

$$S=0.43\bullet erfc(0.0008\,(\Psi\bullet e^{-0.64D}-0.54))+0.27\bullet erfc(0.42\,(\Psi\bullet e^{-0.16D}-0.0003)) \quad (10.2)$$

The $LD_{90}$ for the combined experimental works is 1,083 J cm$^{-2}$, compared with only 100.8 J cm$^{-2}$ for the E. coli experiment. This wide variability of bacterial susceptibility

**Table 10.2:** Soil bacterial numbers as a function of microwave treatment, soil depth and recovery time after treatment (Entries in the table with different superscripts are significantly different to one another) (Source: Brodie et al., 2015).

| Soil Depth (cm) | Time from Microwave Treatment (Days) | Estimated Microwave Treatment ($J\ cm^{-2}$) 0 | 150 | 300 | 600 |
|---|---|---|---|---|---|
| 0 | 1 | 6.20[d] | 5.57[d] | 4.73[d] | 1.78[d] |
| | 31 | 18.90[c] | 38.48[a] | 38.25[a] | 19.67[c] |
| 5 | 1 | 3.78[d] | 4.71[d] | 4.23[d] | 1.18[d] |
| | 31 | 18.73[c] | 24.28[bc] | 29.95[b] | 28.22[b] |
| 10 | 1 | 4.06[d] | 2.93[d] | 3.87[d] | 1.74[d] |
| | 31 | 16.93[c] | 26.13[bc] | 28.90[b] | 18.00[c] |
| LSD (P = 0.05) | | | | | 7.30 |

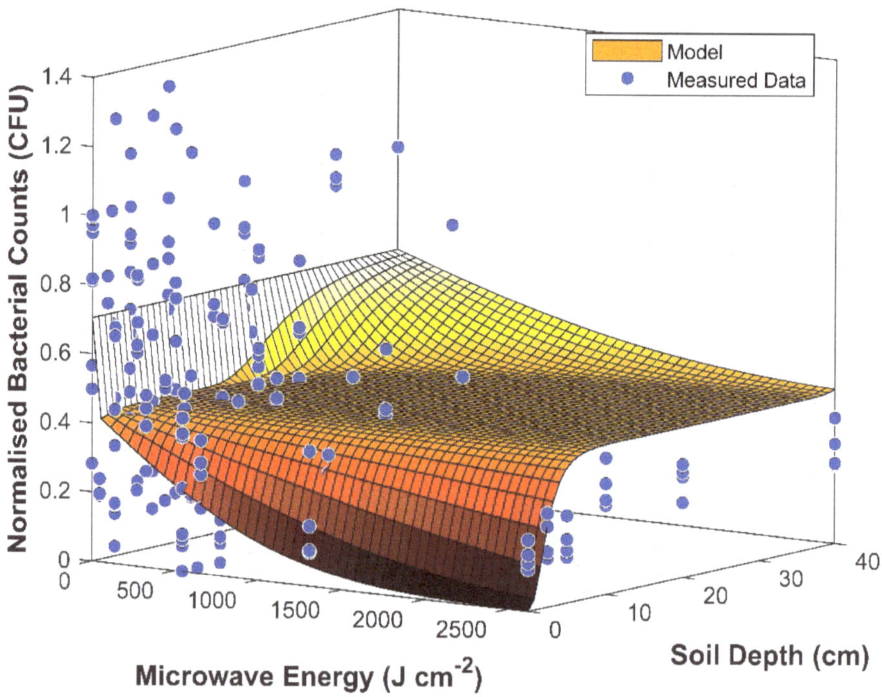


**Figure 10.2:** Combined response of bacteria to microwave treatment, as a function of applied energy density and soil depth (Sources of some data: Brodie et al., 2015; Cooper & Brodie, 2009).

to microwave treatment is apparent from other literature as well. In a review by Shamis, Croft, Taube, Crawford, and Ivanova (2012), they refer to microwave radiation (at 45°C and a frequency of 18 GHz) being used to sterilise transplant biomaterial of pathogenic bacteria (*E. coli* and *S. aureus*) without compromising tissue functionality and durability of the transplant materials. According to their study, fatality in the bacteria was achieved at 45°C. On the other hand, Vela et al. (1976) found that nitrifying bacteria were resistant for 40,000 joules of microwave energy, at 2.45 GHz applied in a modified microwave oven cavity, when the soil temperature was in excess of 80°C.

Analysis of soil biota data gathered from field experiments using the Zapper III prototype developed by Wayland, Davis, and Merkle (1978), presented in Vela et al. (1976), revealed no statistically significant effects of soil condition (wet or dry), microwave energy (0, 200, 400, and 800 J $cm^{-2}$), or soil depth (0-2.5, 2.5-5.0, and 5.0-10.0 cm) on soil biota, except in the case of fungi's response to microwave energy and soil depth (separately). Their data shows a 40% reduction in fungal populations, compared with the control, in response to microwave energy (Figure 10.3). It also showed that fungi in the top few centimetres of soil were more affected than those at deeper depths (Figure 10.4). These are consistent with the heating effects of microwave energy in soil.

## 10.3 Assessment of Fungi and other Soil Biota

To better understand the potential effect of microwave soil treatment on soil biota, other than bacterial, Brodie et al. (2015) undertook fluorescence analyses of the soil samples used to assess the bacterial populations during the Fluorescence microscopy experiment described above. In addition to bacterial, they also considered total fungi, Flagellates, Amoeba, and Ciliates.

These biotas experienced no statistically significant effect that could be attributed to microwave treatment dose response (Tab. 10.3 to Tab. 10.6), even though there were some significant results in the data. Although every effort was made to randomise the soil sampling procedure, it is likely that significant differences in these data were due to natural spatial variability of soil biota captured in the soil sampling process.

The Fluorescence microscopy protocol considers all species of fungi found in the soil and does not address the impact of microwave soil treatment at a species level. In one further experiment, two pathogenic fungal species (*Fusarium oxysporum* and *Sclerotium rolfsii*) were investigated. The experiment took place at a field site in summer of 2016. The microwave trailer prototype system, described in Chapter 8, was set up in the field to slowly move over the soil during the experiment. Samples of barley grain were inoculated with *Fusarium oxysporum*, placed in porous nylon bags and buried at depths of 2.5 cm, 5.0 cm and 10.0 cm in the soil in front of the moving trailer. Sclerotia of *Sclerotium rolfsii* were also placed into nylon bags and buried at the same depths in front of the moving microwave trailer. Control samples of both species

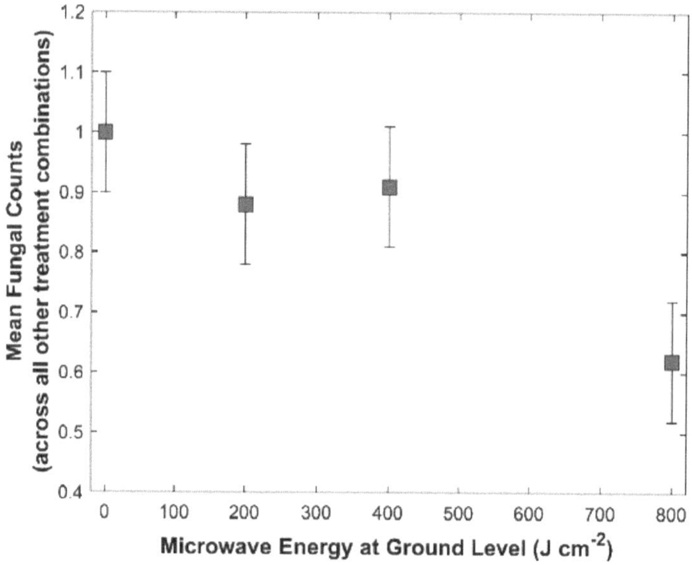


**Figure 10.3:** Mean fungal counts across all treatment combinations in experiments by Vela et al. (1976). Error bars indicate Least Significant Differences (P = 0.05).

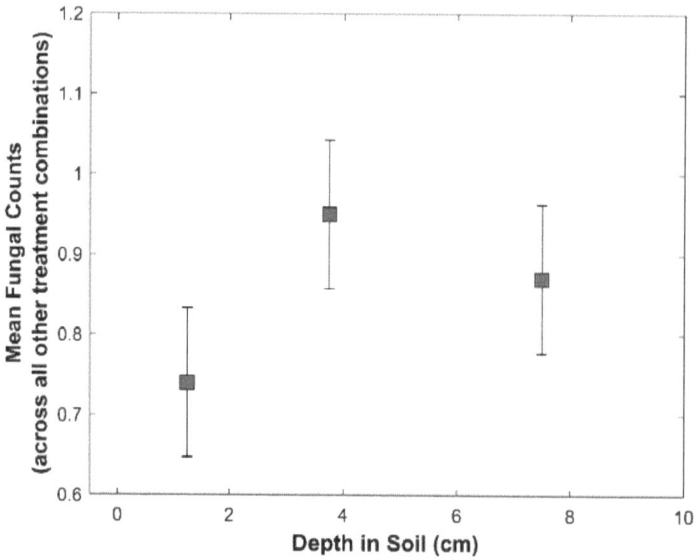


**Figure 10.4:** Mean fungal counts across all treatment combinations, as a function of soil depth, in experiments by Vela et al. (1976). Error bars indicate Least Significant Differences (P = 0.05).

**Table 10.3:** Total fungal numbers as a function of microwave treatment, soil depth and recovery time after treatment (Entries in the table with different superscripts are significantly different to one another) (Source: Brodie et al., 2015).

| Soil Depth (cm) | Time from Microwave Treatment (Days) | Estimated Microwave Treatment ($J\ cm^{-2}$) | | | |
|---|---|---|---|---|---|
| | | 0 | 150 | 300 | 600 |
| 0 | 1 | 79.03[a] | 850.73[a] | 220.98[a] | 240.18[a] |
| | 31 | 209.48[a] | 146.13[a] | 191.00[a] | 253.88[a] |
| 5 | 1 | 77.30[a] | 4443.50[b] | 185.65[a] | 108.43[a] |
| | 31 | 146.90[a] | 142.45[a] | 171.10[a] | 223.86[a] |
| 10 | 1 | 36.83[a] | 380.95[a] | 106.48[a] | 114.66[a] |
| | 31 | 106.33[a] | 76.30[a] | 150.55[a] | 133.08[a] |
| LSD (P = 0.05) | | | | | 1956.60 |

**Table 10.4:** Flagellate numbers as a function of microwave treatment, soil depth and recovery time after treatment (Entries in the table with different superscripts are significantly different to one another) (Source: Brodie et al., 2015).

| Soil Depth (cm) | Time from Microwave Treatment (Days) | Estimated Microwave Treatment ($J\ cm^{-2}$) | | | |
|---|---|---|---|---|---|
| | | 0 | 150 | 300 | 600 |
| 0 | 1 | 4311.00[a] | 2931.50[b] | 2167.00[b] | 1855.38[b] |
| | 31 | 1208.50[b] | 4000.75[a] | 397.33[b] | 1536.88[b] |
| 5 | 1 | 2567.50[b] | 3343.25[b] | 2303.50[b] | 2672.50[b] |
| | 31 | 1386.75[b] | 1414.50[b] | 1068.33[b] | 499.75[b] |
| 10 | 1 | 1902.25[b] | 310.75[b] | 469.00[b] | 1901.13[b] |
| | 35 | 965.00[b] | 1282.25[b] | 246.75[b] | 184.75[b] |
| LSD (P = 0.05) | | | | | 2654.23 |

**Table 10.5:** Amoeba numbers as a function of microwave treatment, soil depth and recovery time after treatment (Entries in the table with different superscripts are significantly different to one another) (Source: Brodie et al., 2015).

| Soil Depth (cm) | Time from Microwave Treatment (Days) | Estimated Microwave Treatment ($J\ cm^{-2}$) | | | |
|---|---|---|---|---|---|
| | | 0 | 150 | 300 | 600 |
| 0 | 1 | 2859.50[a] | 29406.25[b] | 1889.00[a] | 7563.75[a] |
| | 31 | 2299.75[a] | 5722.75[a] | 2626.67[a] | 2458.50[a] |
| 5 | 1 | 941.50[a] | 6411.00[a] | 1303.50[a] | 10862.63[a] |
| | 31 | 2076.25[a] | 3785.25[a] | 1809.33[a] | 2280.63[a] |
| 10 | 1 | 926.50[a] | 4956.25[a] | 1037.50[a] | 4431.25[a] |
| | 31 | 735.25[a] | 2191.75[a] | 287.75[a] | 1179.25[a] |
| LSD (P = 0.05) | | | | | 10653.90 |

**Table 10.6:** Ciliate numbers as a function of microwave treatment, soil depth and recovery time after treatment (Entries in the table with different superscripts are significantly different to one another) (Source: Brodie et al., 2015).

| Soil Depth (cm) | Time from Microwave Treatment (Days) | Estimated Microwave Treatment ($J\ cm^{-2}$) | | | |
|---|---|---|---|---|---|
| | | 0 | 150 | 300 | 600 |
| 0 | 1 | 650.00[a] | 1015.25[a] | 119.75[a] | 196.38[a] |
| | 31 | 1747.25[a] | 505.25[a] | 115.67[a] | 88.13[a] |
| 5 | 1 | 45.25[a] | 1908.50[b] | 50.25[a] | 573.13[a] |
| | 31 | 91.50[a] | 127.00[a] | 46.33[a] | 241.75[a] |
| 10 | 1 | 403.75[a] | 37.50[a] | 41.25[a] | 53.62[a] |
| | 31 | 127.00[a] | 109.25[a] | 123.25[a] | 83.13[a] |
| LSD (P = 0.05) | | | | | 1132.89 |

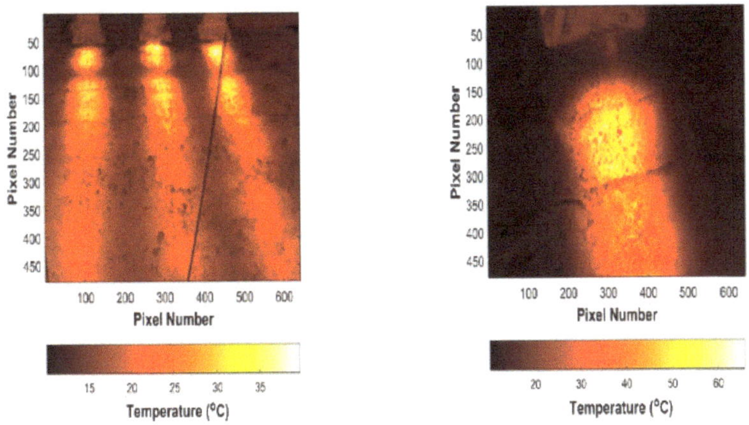


**Figure 10.5:** Thermal images of the soil surface during microwave treatment using the prototype trailer.

were also buried in the soil about 2 m from the path of the trailer. All experimental protocols were replicated five times for each species. The applied microwave energy was approximately 2,300 J $cm^{-2}$ along the treated strips of soil.

A thermal camera was used to evaluate the surface soil heating from the microwave trailer as it moved very slowly (approximately 4 metres per hour) over the soil surface. Nylon sample bags were extracted from the soil 60 minutes after treatment and the fungi were grown out on nutrient media to evaluate their survival and growth behaviour.

Figure 10.5 shows thermal images of the soil surface temperature during microwave treatment of the soil at the Toolangi experimental site. Based on previous experimental work (Brodie et al., 2007), the maximum soil temperature is about 2 to 3 cm below the surface. It is also apparent that the soil remains hot for some time after treatment. observations indicated that about 30 minutes was required for the surface temperature to return to ambient temperature; however, cooling rate diminishes exponentially with time.

Table 10.7 summarises the results for the fungal experiment. *Fusarium oxysporum* is more susceptible to microwave treatment than *Sclerotium rolfsii*; however, in both cases the surviving fungi from the microwave treated samples exhibited significantly suppressed growth, which indicates that sub-lethal microwave treatment may provide some benefit was well.

Other experiments (unpublished) have demonstrated efficacy of microwave soil treatment, *in situ* at between 300 and 400 J $cm^{-2}$, on:

1. Nematodes - *Sarpofage aaltjes*, *Tylenchorhynchus* spp., and *Helicotylenchus* spp.
2. Fungi - *Alternaria* spp., *Fusarium oxysporum*, *Fusarium solani*, *Fusarium* spp., *Phytophthora capsica*, *Phytophthora* spp., and *Verticillium* spp.

**Table 10.7:** Mortality of two pathogenic fungi as a function of burial depth and microwave treatment.

| Species | Burial Depth (cm) | Survival Fraction |
|---|---|---|
| *Fusarium oxysporum* | Control | 1.0 |
| | 2.5 | 0.19 |
| | 5.0 | 0.66 |
| | 10.0 | 0.75 |
| *Sclerotium rolfsii* | Control | 1.0 |
| | 2.5 | 0.39 |
| | 5.0 | 0.36 |
| | 10.0 | 0.62 |

Microwave treatment of *Ascochyta rabiei* spores have also been successful, with 100% mortality when the treated medium was heated above 70°C.

## 10.4 Some Thoughts about Soil Biota

This research has revealed that some species of bacteria are particularly susceptible to microwave treatment, with the general population of fungi and protozoa being less affected (Brodie et al., 2015); however, more recent research has revealed that some species of fungi appear to be more susceptible to microwave treatment than others. Nelson (1996) points out that any effect of microwave treatment on soil born objects is probably associated with the heating effect on the bulk soil. The fatal impacts of high temperatures on plants and other organisms have been studied in detail for over a century (Levitt, 1980). Various models have been developed, which relate temperature and time to survival rates of various organisms.

A thoroughly demonstrated empirical relationship between lethal temperature and temperature holding time for plants and plant parts (including seeds) has been developed by Lepeschkin (1912):

$$T=79.8\text{-}12.8\bullet log_{10} Z \qquad (10.3)$$

Where T is the lethal temperature (°C), and Z is the lethal temperature holding time, in minutes (Levitt, 1980).

Studies by Shlevin et al. (Shlevin, Saguy, Mahrer, & Katan, 2003) investigated the impact of temperature and holding time on the pathogenic fungi, *Fusarium oxysporum* and *Sclerotium rolfsii*, in soil. They derived a Weibull distribution from their data:

$$S=e^{-bt^n} \tag{10.4}$$

Where S is the survival fraction of the fungal population, b is a parameter associated with each species and temperature, t is holding time (hours), and n is another species-specific parameter.

Studies by Trevisani, Mancusi, and Valero (2014), investigating the effect of temperature on *E. coli* bacteria, indicate that the lethality can be described by:

$$S=e^{-bt} \tag{10.5}$$

Where b is a function of temperature.

Studies by Noling (1997), investigating the effect of temperature on the Southern root knot nematode (*Meloidogyne incognita*), also indicate that the lethality can be described by:

$$S=10^{-bt} \tag{10.6}$$

Where b is a function of temperature.

Figure 10.6 illustrates the relationship between fatal temperature and holding time, based on equations (10.3) to (10.6). Although these empirical models are linked to conventional heating methods, they agree with the key findings outlined in this research. Nematodes are the most vulnerable to microwave soil treatment, followed by some bacteria species. Some species of are more vulnerable to microwave soil treatment while other fungi have the potential to survive much higher microwave energy doses.

It should also be noted that fatality of a treatment is not just a function of temperature, but of holding time as well. For example, treatment of fungal samples in soil had a much more evident effect on their populations than treatment in small pots (Brodie et al., 2015), because the soil in these pots may have cooled too quickly to achieve the necessary temperature-time requirement for fatality. It is now surmised that rapid cooling of the pots during the earlier experiment (Brodie et al., 2015) may have negated the temperature effects on total fungi and soil protozoa.

Although some insights into the responses of soil biota have been determined, there are still many uncertainties about the effect of microwave soil treatment on these species. Clearly, soil biota also affects plant growth, so the next chapter will explore the effect of microwave soil treatment on crop plant performance.

## 10.5 Conclusions

Microwave treatment reduces bacterial populations in the top layers of soil, but populations that are deeper in the soil are relatively unaffected. Bacterial populations

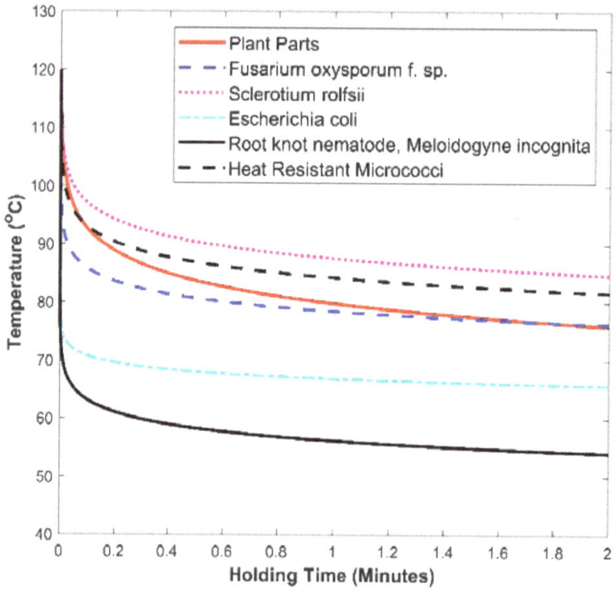


**Figure 10.6:** Fatal Temperature – Time curves for: Root knot nematodes, two species of bacteria, plant parts, and two species of fungi.

increased significantly within a month of microwave treatment. *E. coli* populations experienced a $10^{-5}$ reduction in numbers in the top layer of soil when 500 J cm$^{-2}$ of microwave energy was applied to the surface; however other soil bacteria survived over 2000 J cm$^{-2}$ of microwave energy applied to the soil surface, suggesting that some species are more susceptible to microwave treatment than others. The impact of microwave treatment on soil fungi is evident; however, in practice it seems that the soil temperature must be maintained for some time to be effective.

## 10.6 References


Bebawi, F. F., Cooper, A. P., Brodie, G. I., Madigan, B. A., Vitelli, J. S., Worsley, K. J., & Davis, K. M. (2007). Effect of microwave radiation on seed mortality of rubber vine (Cryptostegia grandiflora R.Br.), parthenium (Parthenium hysterophorous L.) and bellyache bush (Jatropha gossypiifolia L.). Plant Protection Quarterly, 22(4), 136-142.

Brodie, G., Grixti, M., Hollins, E., Cooper, A., Li, T., & Cole, M. (2015). Assessing the Impact of Microwave Treatment on Soil Microbial Populations. Global Journal of Agricultural Innovation, Research & Development, 2(1), 25-32.

Brodie, G., Hamilton, S., & Woodworth, J. (2007). An assessment of microwave soil pasteurization for killing seeds and weeds. Plant Protection Quarterly, 22(4), 143-149.

Brodie, G., Harris, G., Pasma, L., Travers, A., Leyson, D., Lancaster, C., & Woodworth, J. (2009). Microwave soil heating for controlling ryegrass seed germination. Transactions of the American Society of Agricultural and Biological Engineers, 52(1), 295-302.

Brodie, G., Harris, G., & Torgovnikov, G. (2014). Microwave Control of Weed Seeds in Biosolids. Paper presented at the 19th Australasian Weed Conference, Hobart, Australia.

Brodie, G., & Hollins, E. (2015). The Effect of Microwave Treatment on Ryegrass and Wild Radish Plants and Seeds. Global Journal of Agricultural Innovation, Research & Development, 2(1), 16-24.

Brodie, G., Ryan, C., & Lancaster, C. (2012a). The effect of microwave radiation on Paddy Melon (Cucumis myriocarpus). International Journal of Agronomy, 2012, 1-10.

Brodie, G., Ryan, C., & Lancaster, C. (2012b). Microwave technologies as part of an integrated weed management strategy: A Review. International Journal of Agronomy, 2012, 1-14.

Cooper, A. P., & Brodie, G. (2009). The effect of microwave radiation and soil depth on soil pH, N, P, K, SO4 and bacterial colonies. Plant Protection Quarterly, 24(2), 67-70.

Cornea, A., & Conn, P. M. (2014). Fluorescence Microscopy. [electronic resource] : Super-Resolution and other Novel Techniques: Burlington : Elsevier Science, 2014.

Davis, F. (1975). “Zapper” blasts weed seeds. New Zealand Journal of Agriculture.

Davis, F. S. (1974). New techniques in weed control via microwaves. Paper presented at the Southern Nurserymen’s Association Conference, Nacogdoches, Texas, U.S.A.

Davis, F. S., Wayland, J. R., & Merkle, M. G. (1971). Ultrahigh-Frequency Electromagnetic Fields for Weed Control: Phytotoxicity and Selectivity. Science, 173(3996), 535-537.

Davis, F. S., Wayland, J. R. and Merkle, M. G. (1973). Phytotoxicity of a UHF Electromagnetic Field. Nature, 241(5387), 291-292.

Devine, A. A., Grunden, A. M., Krisiunas, E., Davis, D. K., Rosario, G., Scott, S., et al. (2007). Testing the Efficacy of a Combination of Microwave and Steam Heat for Log Reduction of the Microbial Load Following a Simulated Poultry Mass Mortality Event. Applied Biosafety, 12(2), 79-84.

Lepeschkin, W. W. (1912). Zur Kenntnis der Einwirkung supamaximaler Temperaturen auf die Pflanze. Berichte der Deutschen Botanishen Gesellschaft, 30, 713-714.

Levitt, J. (1980). Response of Plants to Environmental Stresses (Vol. 1). New York: Academic Press.

Malarić, K., Štambuk, A., Šrut, M., & Tkalec, M. (2008). Evaluation of genotoxic potential of radiofrequency/microwave electromagnetic field (RF/MW EMF) using comet assay in earthworms (Eisenia fetida). Paper presented at the 16th IMEKO TC4 Int. Symp.: Exploring New Frontiers of Instrum. and Methods for Electrical and Electronic Measurements; 13th TC21 Int. Workshop on ADC Modelling and Testing - Joint Session, Proc.

Nelson, S. O. (1996). A review and assessment of microwave energy for soil treatment to control pests. Transactions of the ASAE, 39(1), 281-289.

Noling, J. W. (1997). Relative lethal dose, a Time-Temperature model for relating soil solarization efficacy and treatment duration for nematode control. Retrieved from http://www.crec.ifas.ufl.edu/extension/soilipm/1997%20MBAO/J.%20Noling/J.%20Noling%20(17)%201997%20Presentation.pdf

Rahi, G. S., & Rich, J. R. (2008). Potential of microwaves to control plant-parasitic nematodes in soil. Journal of Microwave Power & Electromagnetic Energy, 42(1), 5-42112.

Rahi, G. S., & Rich, J. R. (2011). Effect of Moisture on Efficiency of Microwaves to Control Plant - Parasitic Nematodes in Soil. Journal of Microwave Power and Electromagnetic Energy, 45(2), 86-93.

Samtani, J. B., Gilbert, C., Ben Weber, J., Subbarao, K. V., Goodhue, R. E., & Fennimore, S. A. (2012). Effect of Steam and Solarization Treatments on Pest Control, Strawberry Yield, and Economic Returns Relative to Methyl Bromide Fumigation. HortScience, 47(1), 64-70.

Shamis, Y., Croft, R., Taube, A., Crawford, R., & Ivanova, E. (2012). Review of the specific effects of microwave radiation on bacterial cells. Applied Microbiology and Biotechnology, 96(2), 319-325. doi:10.1007/s00253-012-4339-y

Shlevin, E., Saguy, I. S., Mahrer, Y., & Katan, J. (2003). Modeling the Survival of Two Soilborne Pathogens Under Dry Structural Solarization. Phytopathology, 93(10), 1247-1257. doi:10.1094/PHYTO.2003.93.10.1247

Speir, T. W., Cowling, J. C., Sparling, G. P., West, A. W., & Corderoy, D. M. (1986). Effects of microwave radiation on the microbial biomass, phosphatase activity and levels of extractable N and P in a low fertility soil under pasture. Soil Biology and Biochemistry, 18(4), 377-382. doi:Doi: 10.1016/0038-0717(86)90041-6

Trevisani, M., Mancusi, R., & Valero, A. (2014). Thermal inactivation kinetics of shiga toxin-producing Escherichia coli in buffalo mozzarella curd. Journal of Dairy Science, 97(2), 642-650.

Vela, G. R., Wu, J. F., & Smith, D. (1976). Effect of 2450 MHz microwave radiation on some soil microorganisms in situ. Soil Science, 121(1), 44-51.

Wainwright, M., Killham, K., & Diprose, M. F. (1980). Effects of 2450MHz microwave radiation on nitrification, respiration and s-oxidation in soil. Soil Biology and Biochemistry, 12, 489-493.

Wayland, J. R., Davis, F. S., & Merkle, M. G. (1978). United States Patent No.: C. S. Phytox Corporation, Texas.

# 11 The Effect of Microwave Soil Treatment on Subsequent Crop Growth and Yield

## 11.1 Introduction

Chemical soil fumigation, as a means of controlling weeds, improving crop yields and controlling pathogenic organisms in food and fibre production systems, is commonly practiced in the agricultural industry, especially in high value horticultural crops. Several fumigant chemicals, including Methyl bromide and Metham sodium, have serious health and environmental concerns, which have led to these chemicals being removed from the market (Carter, Chalfant, Goodhue, Han, & DeSantis, 2005; Fennimore, Haar, & Ajwa, 2003; Sances & Ingham, 1997; Sydorovych et al., 2006).
A sustained research programme has demonstrated that microwave treatment of *in-situ* soil, using a horn antenna applicator, can effectively kill weed plants and their seeds (Brodie, Hamilton, & Woodworth, 2007; Brodie, Hollins, & Woods-Casey, 2014; Brodie, Ryan, & Lancaster, 2012). It has also been demonstrated that microwave soil treatment can reduce populations of some organisms in the soil, such as *Escherichia coli* (Brodie, Grixti, et al., 2015), and nematodes (Diprose, Benson, & Willis, 1984); however, the effect of microwave soil treatment on subsequent crop growth has not been studied. This chapter presents the results of recent pot and field experiments to explore the effect of microwave soil treatment on the growth and yield of crop plants grown in the treated soil.

## 11.2 Initial Pot Trials

Several pot trials were initially undertaken to evaluate the impact of microwave soil treatment on the subsequent growth and health of crops. In all cases the soil was allowed to cool to ambient temperature after microwave treatment, before crop seeds were planted into the treated soil. In one of the first experiments, wheat, canola and rice, which are important winter crops in Australia, were used as trial species. Soil was collected from a regularly cropped paddock at Dookie agricultural campus of the University of Melbourne (36° 23' S and 145° 43' E). The soil was placed into 15 cm diameter pots and randomly allocated to 0, 168, 384, and 576 J $cm^{-2}$ microwave treatments using a horn antenna with aperture dimensions of 110 mm by 55 mm, fed from a 2 kW, 2.45 GHz microwave generator. It was assumed that an entirely weed free environment would represent the most ideal growing conditions for these crops, so a hand weeded control was included in the experiment as well.

After cooling overnight, wheat (*Triticum spp.*) and canola (*Brassica napus*), and pre-germinated seeds of rice (*Oryza spp.*) were planted into the pots. The pots of wheat and canola were placed in a glass house at Dookie campus of the University of

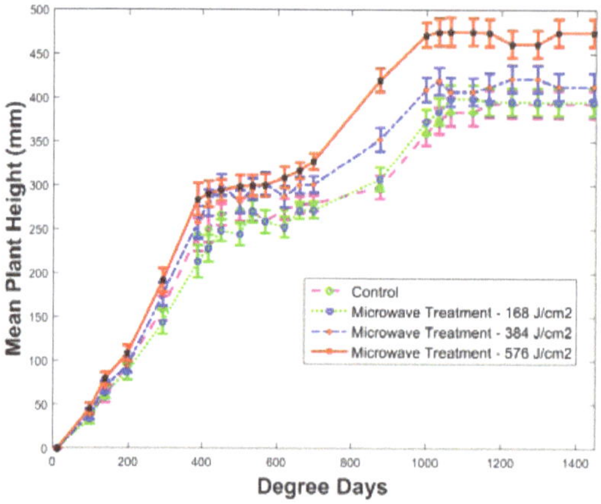


**Figure 11.1:** Mean wheat plant height as a function of time since planting and microwave soil treatment energy (error bars represent LSD for P = 0.05) (Modified from: Brodie, Bootes, & Reid, 2015).

**Figure 11.2:** Comparison of wheat and canola plant growth as a function of microwave treatment energy (Control on the left and highest treatment on the right).

Melbourne and watered three times per week. The rice pots were randomly placed in a flooded rice crop of a commercial rice producer, near the township of Swan Hill in the southern part of Australia (35° 20‘ S and 143° 33‘ E).

After the wheat and canola plants were well established, the pots were thinned to a maximum of three plants per pot. The three most vigorous plants were kept in each pot. Wheat plant heights were monitored through the growing season. Days from planting to flowering was monitored for the Canola plants, as an indicator of crop maturation rate. Final grain yield from all species was assessed.

Plant maturation rate, mean plant height (Figure 11.1 and Figure 11.2), plant/tiller density (Figure 11.3), and mean yield per pot (Tab. 11.1) all increased significantly as the level of applied microwave energy increased.

**Figure 11.3:** Comparison of rice plant growth as a function of microwave treatment energy (Control on the left and highest microwave treatment on the right).

**Table 11.1:** Effect of microwave soil treatment on crop yield and maturation rate (Modified from: Brodie, Bootes, et al., 2015).

| Microwave Treatment (J cm$^{-2}$) | Un-weeded Control | Hand Weeded Control | 168 | 384 | 576 | LSD (P = 0.05) | Change from Hand Weeded Control |
|---|---|---|---|---|---|---|---|
| Canola Dry Pod Yield (g pot$^{-1}$) | 0.27$^{a}$ | 0.56$^{a}$ | 0.36$^{a}$ | 1.25$^{b}$ | 1.95$^{c}$ | 0.55 | 248 % |
| Days to Flowering - Canola | 71.4$^{a}$ | 67.6$^{ab}$ | 70.2$^{a}$ | 63.2$^{b}$ | 61$^{b}$ | 7.1 | 10.8% |
| Wheat Dry Grain Yield (g pot$^{-1}$) | 0.66$^{a}$ | 0.67$^{a}$ | 0.68$^{a}$ | 0.75$^{a}$ | 1.25$^{b}$ | 0.30 | 86.6 % |
| Rice Dry Grain Yield (g pot$^{-1}$) | 40.00$^{a}$ | 41.3$^{a}$ | 43.25$^{a}$ | 59.00$^{ab}$ | 64.00$^{b}$ | 18.90 | 55 % |

Note: entries with different superscripts across the rows are statistically different from one another
Also note: pots used in rice experiment were larger than for other crops – hence higher yield per pot.

**Table 11.2:** Soil nutrient assessments, taken immediately after microwave treatment.

| Microwave Treatment | Total Carbon | Total Nitrogen | Ammonia N | Nitrate N | Bray 2 Phosphate | Total Phosphorus |
|---|---|---|---|---|---|---|
| $J\ cm^{-2}$ | % | % | $mg\ kg^{-1}$ | $mg\ kg^{-1}$ | $mg\ kg^{-1}$ | $mg\ kg^{-1}$ |
| 0 | 1.41 | 0.14 | 5.23[ab] | 109.28 | 79.01 | 554.40 |
| 168 | 1.75 | 0.14 | 4.09[a] | 78.38 | 75.89 | 543.20 |
| 384 | 1.47 | 0.14 | 10.87[b] | 86.20 | 78.84 | 560.00 |
| 576 | 1.47 | 0.14 | 4.92[a] | 99.20 | 80.68 | 578.80 |
| LSD ($P = 0.05$) | 0.52 | 0.016 | 5.71 | 57.81 | 10.49 | 48.37 |

Note: entries with different superscripts down the columns are statistically different from one another

**Table11.3:** Soil nutrient assessments, taken after wheat harvest, as a function of applied microwave energy.

| Microwave Treatment | Total Carbon | Total Nitrogen | Ammonia N | Nitrate N | Bray 2 Phosphate | Total Phosphorus |
|---|---|---|---|---|---|---|
| $J\ cm^{-2}$ | % | % | $mg\ kg^{-1}$ | $mg\ kg^{-1}$ | $mg\ kg^{-1}$ | $mg\ kg^{-1}$ |
| 0 | 1.36[a] | 0.12[ab] | 4.05[ab] | 0.39[c] | 3.87[a] | 405.60 |
| 168 | 1.46[ab] | 0.13[b] | 4.52[ab] | 0.20[ab] | 4.79[c] | 429.60 |
| 384 | 1.53[b] | 0.13[b] | 4.63[b] | 0.28[b] | 4.62[bc] | 423.80 |
| 576 | 1.27[a] | 0.11[a] | 3.62[a] | 0.19[a] | 4.33[b] | 402.60 |
| LSD ($P = 0.05$) | 0.11 | 0.01 | 0.95 | 0.08 | 0.44 | 31.17 |

Note: entries with different superscripts down the columns are statistically different from one another

### 11.2.1 Soil Analysis

About 5 grams of soil from each wheat pot (before and after the crop was grown) was collected for soil nutrient analysis. Initially, there was no effect on soil nutrition, which could be directly attributable to microwave treatments (Tab. 11.2); however, there were significant differences in the soil nutrient status by the end of the wheat growing process (Tab. 11.3).

## 11.3 Pot Trials on Wheat with Combined Microwave and Nitrogen Treatments

From the preliminary pot trials, it was clear that, along with the changes in soil biota outlined in Chapter 10, enhanced nutrient uptake may be contributing to the improvement in plant growth in microwave treated soil. Therefore, further experiments were devised to explore the interaction of microwave soil treatment with nutrient availability, especially nitrogen.

### 11.3.1 An Initial Microwave-Nitrogen Interaction Experiment

A more sophisticated glasshouse experiment was conducted on Gregory winter Wheat (*Triticum aestivum* L.) to examine the impact of microwave energy on the nitrogen uptake, nitrogen use efficiency, plant growth and yield. The experiment was a factorial experiment comprising two factors (nitrogen application and microwave exposure): four concentrations of nitrogen were applied ($N_0$ = 0 mg $^{15}N\ L^{-1}$, $N_1$ = 50 mg $^{15}N\ L^{-1}$, $N_2$ = 100 mg $^{15}N\ L^{-1}$ and $N_3$ = 150 mg $^{15}N\ L^{-1}$) and two levels of microwave (MW) exposure ($T_0$, $T_1$) were used. There was a total of eight treatment combinations with ten replicates per treatment. The $^{15}N$ isotope was used to trace the movement of applied nitrogen.

Based on the earlier experiment the two microwave treatments were $T_0$, which was an untreated option and $T_1$, which was equivalent to the highest microwave treatment in the previous pot experiment. The pots were then sown with the wheat variety Gregory, at a rate of 20 seeds per pot, and transferred to the glasshouse. Following germination, the pots were thinned to 10 plants per pot. The pots were watered to the water holding capacity of the soil (60%) every 3 days. Plant height was measured using a wooden ruler and chlorophyll content of leaves was measured using a chlorophyll meter (SPAD-502 (Soil-Plant Analysis Development); Spectrum Technologies, Inc. Aurora, Illinois, USA), at 20, 40 and 60 days after sowing. Stem diameter, intermodal distance, spike length, number of grains per spike, 100 grain weight, plant dry matter, grain yield (yield per pot) and grain nitrogen percentage were measured at harvest.

The nitrogen derived from fertilizer was calculated by using the following equation as described by Salamanca-Jimenez, Doane, and Horwath (2017):

$$\text{Nddf}\ (\%) = \frac{\text{atom}\ \%\ {}^{15}\text{Nexcess in plant dry biomass}}{\text{atom}\ \%\ {}^{15}\text{Nexcess in fertilizer}} \times \mathbf{100} \tag{11.1}$$

The results showed that microwave exposure had no apparent effect on the plant height of wheat at 60 days after sowing (DAS); however, at 20 DAS and 40 DAS there was a marked increase in plant height in the microwave treated pots in all nitrogen application levels, compared to the untreated soil. Microwave treatment significantly enhanced the chlorophyll content at all levels of applied nitrogen, at all three measurement dates (20 DAS, 40 DAS and 60 DAS). At harvest, microwave treatment resulted in significant increases in all yield components and plant growth parameters.

The data showed that microwave treatment significantly increased the stem diameter of wheat at all the nitrogen levels compared to non-irradiated soil. The maximum stem thickness (3.2 mm) was measured at 150 mg $L^{-1}$ in the microwave treated pots. The rest of $^{15}N$ concentration in microwave treated soil was statistically similar to each other but significantly different from the non-microwave treated counterpart treatments. The minimum stem diameter (1.7 mm) was recorded in the pots where the treatment combination was 0 mg $L^{-1}$ of $^{15}N$ and 0 sec. of microwave treatment.

The results showed that exposure to 2.45 GHz microwave treatment had a significant effect on wheat inter-nodal elongation. The maximum inter-nodal distance (ID; 98.4 mm) was measured in 100 mg $L^{-1}$ concentration followed by the 50 mg $L^{-1}$ (92.5 mm). The $N_1$ (81.9 mm) and $N_4$ (85.8 mm) were statistically similar to each other but gradually differed from pots where no microwave irradiation was applied. The minimum ID (74.7 mm) was measured in the untreated pots with 0 mg $^{15}N$ $L^{-1}$.

The results from this experiment were quite detailed, but of particular interest, there were obvious differences in plant growth (Figure 11.4) and significant differences in grain yield (Figure 11.5). See the Appendix for more details about the other results.

There is a strong relationship ($R^2 = 0.99$) between the extent of yield increase due to microwave soil treatment and the applied nitrogen (Figure 11.6). The relationship between applied nitrogen and the mean increase in wheat grain yield per pot due to microwave treatment is described by:

$$\frac{Y-Y_\infty}{Y_0-Y_\infty} = e^{-aN} \tag{11.2}$$

And:

$$Y = \frac{Y_m-Y_c}{Y_c} \tag{11.3}$$

Where: $Y_m$ = yield from microwave treated pots at a particular nitrogen application rate; $Y_c$ = yield from non-microwave treated pots at a particular nitrogen application rate; $Y_0$ = Y, when zero additional nitrogen is applied to the pots; and $Y_\infty$ = the limit of Y when the applied nitrogen becomes very large.

**Figure 11.4:** Comparison of reproductive development in response to pre-sowing microwave soil heating in wheat crop at nitrogen treatment $N_2$, in untreated control pot (left) and microwave treated pot (right). Adopted from Khan, Brodie, and Gupta (2016).

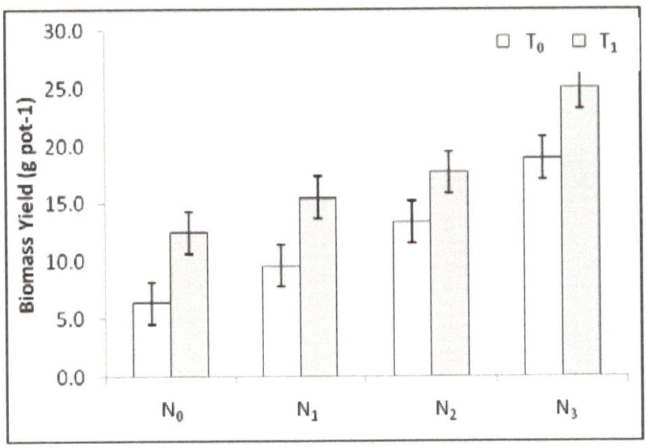


**Figure 11.5:** Biomass yield (g pot-1) of Wheat grown in a glasshouse as a function microwave (MW) treated soil (T0 = untreated (control); T1 = 120 sec 2.45GHz MW treatment under different rates of N application (N0 = 0 mg pot-1; N1= 22.5 mg pot-1; N2 = 45.0 mg pot-1; N3 = 67.5 mg pot-1) measured at harvest. Error bars indicate LSD at 5%.

**Table 11.4:** Summary significance of the results for plant growth and yield parameters measured 20 days after sowing (DAS), 40 DAS, 60 DAS and at harvest; with the p-Values for the analysis of variance (ANOVA) for microwave treatment (MW), nitrogen treatments (N) and the interaction between MW and N. Detailed results are reported in the appendix of this report.

| Parameter | Time of Measurement | p-Values MW | N | MW*N | Figure No. (in Appendix) |
|---|---|---|---|---|---|
| Plant Height (mm) | 20 DAS | <0.001 | <0.001 | 0.003 | Figure 11.20 |
| | 40 DAS | <0.001 | <0.001 | 0.005 | and |
| | 60 DAS | ns | <0.001 | ns | Figure 11.29 |
| Chlorophyll Content | 20 DAS | <0.001 | 0.044 | ns | |
| | 40 DAS | <0.001 | <0.001 | ns | Figure 11.21 |
| | 60 DAS | <0.001 | <0.001 | ns | |
| Grain Yield (g $pot^{-1}$) | Harvest | <0.001 | <0.001 | ns | Figure 11.22 |
| Spikelets per Spike | Harvest | <0.001 | <0.001 | ns | Figure 11.23 |
| Grains per Spikelet | Harvest | <0.001 | <0.001 | 0.032 | Figure 11.24 |
| Biomass yield (g $pot^{-1}$) | Harvest | <0.001 | <0.001 | ns | Figure 11.5 |
| Harvest Index | Harvest | ns | 0.028 | 0.049 | Figure 11.25 |
| Tillers per Shoot | Harvest | <0.001 | <0.001 | <0.001 | Figure 11.26 |
| Inter-nodal distance (mm) | Harvest | <0.001 | 0.007 | ns | Figure 11.27 |
| Stem diameter (mm) | Harvest | <0.001 | <0.001 | ns | Figure 11.28 |

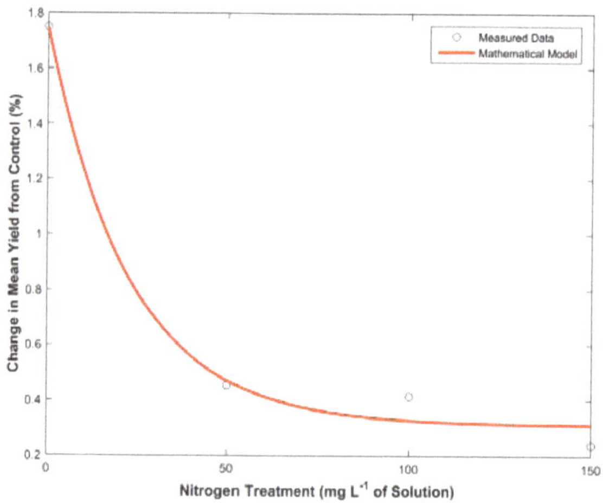


**Figure 11.6:** Change in mean wheat grain yield induced by microwave soil treatment, compared to the equivalent control treatment, as a function of applied nitrogen.

### 11.3.2 A Repeat Microwave-Nitrogen Interaction Experiment, using the Same Soil

The pots with their respective soils, from the first Microwave-Nitrogen Interaction Experiment, were kept in the glasshouse over summer. The soil was reused for a follow-on experiment to evaluate any continuing response to microwave treatment. The soil was not treated in any further way other than Mono-ammonium Phosphate (MAP) was added to all the pots at seed sowing.

Although there was no further microwave treatment of the soil in the pots from the first experiment, microwave treatment significantly affected all measured parameters (Tab. 11.5). The tallest plants (50.92 cm) were in the microwave treated soil to which no nitrogen was applied in the first year of the trial, and the smallest plants were in the untreated soil with 50 mg $L^{-1}$ of $^{15}N$ (41.76 cm) solution added during the first experiment.

Both microwave treatment and nitrogen application individually affected Chlorophyll content of the plants; however, there was no significant interaction of the two factors (Tab. 11.5). Plants with the highest mean Chlorophyll content (44.42) were in the microwave treated soil with no added nitrogen and those with the lowest mean Chlorophyll content (39.19) were in the untreated soil with 150 mg $L^{-1}$ of $^{15}N$ solution added during the first experiment. The treatment with the greatest biomass for the four randomly sampled plants per pot (0.76 g $pot^{-1}$) was the microwave treated soil with the 50 mg $L^{-1}$ of $^{15}N$ solution added in the first experiment and those with the least mean biomass (0.34 g $pot^{-1}$) were in the untreated soil with no added nitrogen (Tab. 11.16 – in the appendix to this chapter).

Plants with the longest flag leaf (23.48 cm) were in the microwave treated soil with no added nitrogen and those with the shortest flag leaf (14.5 cm) were in the non-treated soil with 150 mg $L^{-1}$ of $^{15}N$ solution added. The effect of microwave soil treatment continues beyond a single season (Figure 11.7).

At 85% crop maturity, a small amount (50 gm) of soil was sampled from each pot and immediately shifted on an ice freezer in the laboratory and stored at –20°C until further action. Soil DNA was extracted from a 250 mg sub-sample of these stored soil sample using Power Soil™ DNA Isolation Kit (MoBio Laboratories Inc., Carlsbad, CA, USA). The Mini Bead beater was used for cell lysis at 3000 rpm for 1 minute. All the DNA extracts were quantified using a Nanodrop™ ND2000c spectrophotometer (NanoDrop Technologies, Wilmington, DE, USA) and the quality was ensured with 1% agarose gel electrophoreses.

The *amo*A functional gene of archaea and bacteria were quantified using a CFX96™ optical qPCR detection system (Bio-Rad, Laboratories Inc., Hercules, CA, USA). Each archaeal-*amo*A qPCR reaction of 20 µl contained 10 µl SensiFAST (Bio-Rad Laboratories, USA), 1 µl of each primer set (10 µM; *Arch*-amoAF and *Arch*-amoAR) (Francis, Roberts, Beman, Santoro, & Oakley, 2005), 2 µl of 10-fold diluted DNA template (10 – 115 ng) and 6 µL of water. Each bacterial-*amo*A qPCR reaction of 20 µl contained 10 µl iTaq Universal SYBR GREEN Supermix (Bio-Rad Laboratories, USA),

**Table 11.5:** Summary significance of the results for plant growth in soil conserved from the Previous Experiment; with the p-Values for the analysis of variance (ANOVA) for microwave treatment (MW), nitrogen treatments (N) and the interaction between MW and N. Detailed results are reported in the appendix of this report.

| Parameter | p-Values | | | Table No. |
|---|---|---|---|---|
| | MW | N | MW*N | (in Appendix) |
| Plant Height (cm) | <0.001 | ns | 0.01 | Table 0.14 |
| Chlorophyll Content | 0.003 | <0.001 | 0.01 | Table 0.15 |
| Biomass yield at Tillering (g pot$^{-1}$) | <0.001 | ns | ns | Table 0.16 |
| Flag Leaf Length (cm) | <0.001 | <0.001 | ns | Table 0.17 |
| Biomass yield at Harvest (g pot$^{-1}$) | <0.001 | <0.001 | ns | Table 0.18 |
| Grain yield at Harvest (g pot$^{-1}$) | <0.001 | <0.001 | ns | Table 0.19 |

1 µl of each primer set (10 µM; amoA-1F and amoA-2R) (Rotthauwe, Witzel, & Liesack, 1997), 2 µl of 10-fold diluted DNA template (10 – 115 ng) and 6 µL of water. The thermal cycling conditions for both archaeal and bacterial amoA gene quantification were used: 95°C for 3 min then 40 cycles of 95°C for 5 sec, 60°C for 30 sec and 72°C for 45 sec. For all qPCR reactions, the efficiency was 85 – 100%.

The microwave treatment of soil did not affect an abundance of ammonia oxidizer archaea (Figure 11.8) and bacteria (Figure 11.9) under all level of nitrogen application, which reflects the most likely recovery of nitrifying bacteria after 300 days of MW irradiation if any reduction was induced by microwave energy. The resistance of these beneficial microorganisms had already been reported elsewhere up to microwave energy level of 40,000 J cm$^{-2}$.

### 11.3.3 A Repeat Microwave-Nitrogen Interaction Experiment, using Fresh Soil

The experimental protocol was the same as for the original Microwave-Nitrogen Interaction experiment outlined earlier. Fresh soil was harvested from an adjacent paddock to where the original soil was harvested and subjected to the same treatments. The paddock from which the new soil was harvested, was the same soil type, but had a recent history of being cropped with wheat. Three randomly selected plants were harvested from each pot, dried and weighed. Plant height was measured. Chlorophyll content has been measured using a SPAD meter and photosynthetic rate has been measured using a Licor-6400 system.

As in the previous experiments, microwave treatment significantly affected all measured parameters (Tab. 11.6). The tallest plants (37.59 cm) were in the microwave treated soil to which 100 mg L$^{-1}$ of $^{15}N$ solution added and the shortest plants (32.41

**Figure 11.7:** Comparison of microwave plants growing in soil that was treated with microwave energy at the beginning of the previous wheat crop (three pots on the left) and plants growing in soil from the previous crop that was not treated with microwave energy (three pots on the right).

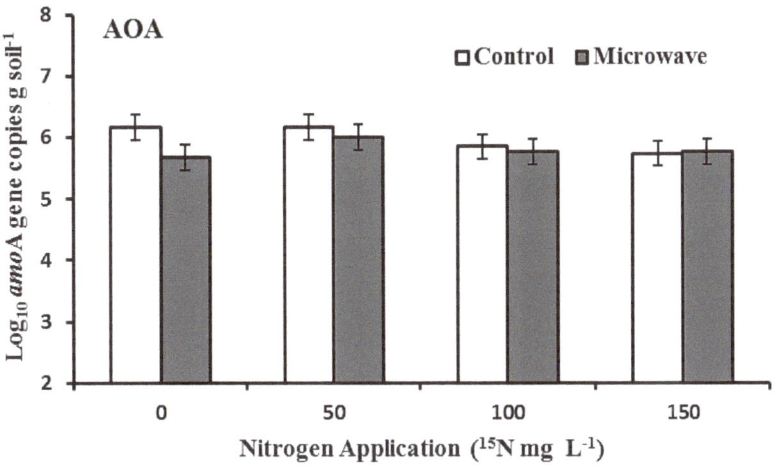


**Figure 11.8:** Effect of microwave soil heating on an abundance of ammonia oxidizer archaea after 300 days of treatment.

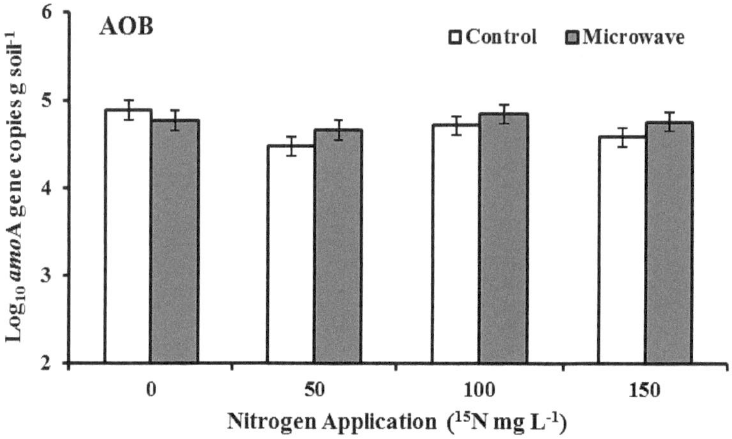


**Figure 11.9:** Effect of microwave soil heating on an abundance of ammonia oxidizer bacteria after 300 days of treatment.

cm) were in the non-treated soil to which 150 mg $L^{-1}$ of $^{15}N$ solution added (Tab. 11.20). Microwave treatment significantly affected Chlorophyll content, but applied nitrogen had no significant effect (Tab. 11.6). Plants with the highest mean Chlorophyll content (49.94) were in the microwave treated soil to which 150 mg $L^{-1}$ of $^{15}N$ solution added and those with the lowest mean Chlorophyll content (42.97) were in the untreated soil with 50 mg $L^{-1}$ of $^{15}N$ solution added (Tab. 11.21). The treatment with the greatest biomass for the three randomly sampled plants per pot (3.54 g $pot^{-1}$) was the microwave treated soil with 150 mg $L^{-1}$ of $^{15}N$ solution added and those with the least mean biomass (2.15 g $pot^{-1}$) were in the untreated soil with 50 mg $L^{-1}$ of $^{15}N$ solution added (Tab. 11.22). Microwave treatment significantly affected photosynthetic rate in the wheat plants (Tab. 11.6). Plants with the highest photosynthetic rate (18.73 μmole $CO_2$ $m^{-2}$ $s^{-1}$) were in the microwave treated soil to which 100 mg $L^{-1}$ of $^{15}N$ solution was added while those with the lowest photosynthetic rate (10.19 μmole $CO_2$ $m^{-2}$ $s^{-1}$) were in the untreated soil to which 150 mg $L^{-1}$ of $^{15}N$ solution (Tab. 11.23).

By using $^{15}N$, it is possible to trace the fate of the fertiliser in the experiment. Figure 11.10 demonstrates that, across both freshly treated soil experiments (2015 and 2016), the plants growing in the microwave treated soil extract less nitrogen from the applied fertiliser than the plants growing in the untreated soil; however, the plants growing in the microwave treated soil yielded more (Figure 11.11).

**Table 11.6:** Summary significance of the results for plant growth in soil conserved from Experiment 1; with the pValues for the analysis of variance (ANOVA) for microwave treatment (MW), nitrogen treatments (N) and the interaction between MW and N. Detailed results are reported in the appendix of this report.

| Parameter | pValues | | | Table No. |
|---|---|---|---|---|
| | MW | N | MW*N | (in Appendix) |
| Plant Height (cm) | <0.001 | 0.05 | ns | Table 0.20 |
| Chlorophyll Content | <0.001 | ns | 0.01 | Table 0.21 |
| Biomass yield at Tillering (g $pot^{-1}$) | <0.001 | ns | ns | Table 0.22 |
| Photosynthetic rate (μmole $CO_2$ $m^{-2}$ $s^{-1}$) | <0.001 | 0.035 | ns | Table 0.23 |
| Biomass yield at Harvest (g $pot^{-1}$) | <0.001 | 0.018 | ns | Table 0.24 |
| Grain yield at Harvest (g $pot^{-1}$) | <0.001 | 0.02 | ns | Table 0.25 |

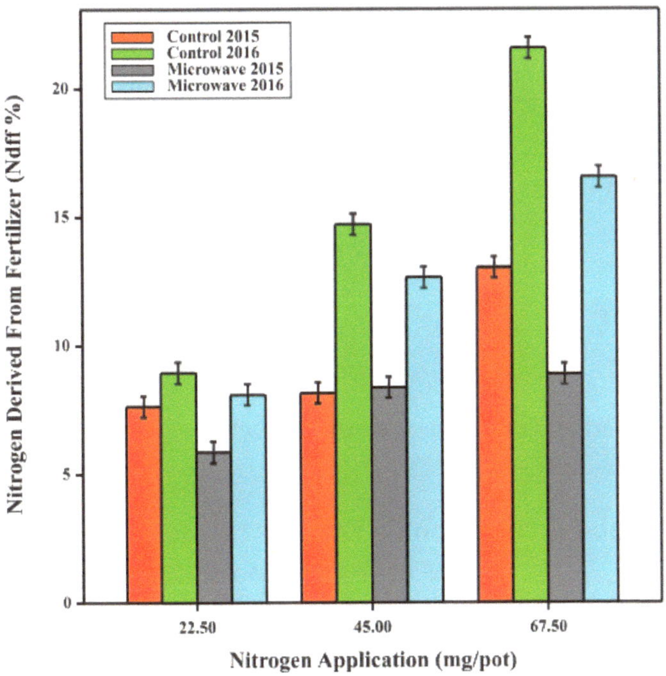


**Figure 11.10:** Effect of microwave soil heating on nitrogen derived from applied fertilizer in wheat crop under two years of microwave-nitrogen experiments.

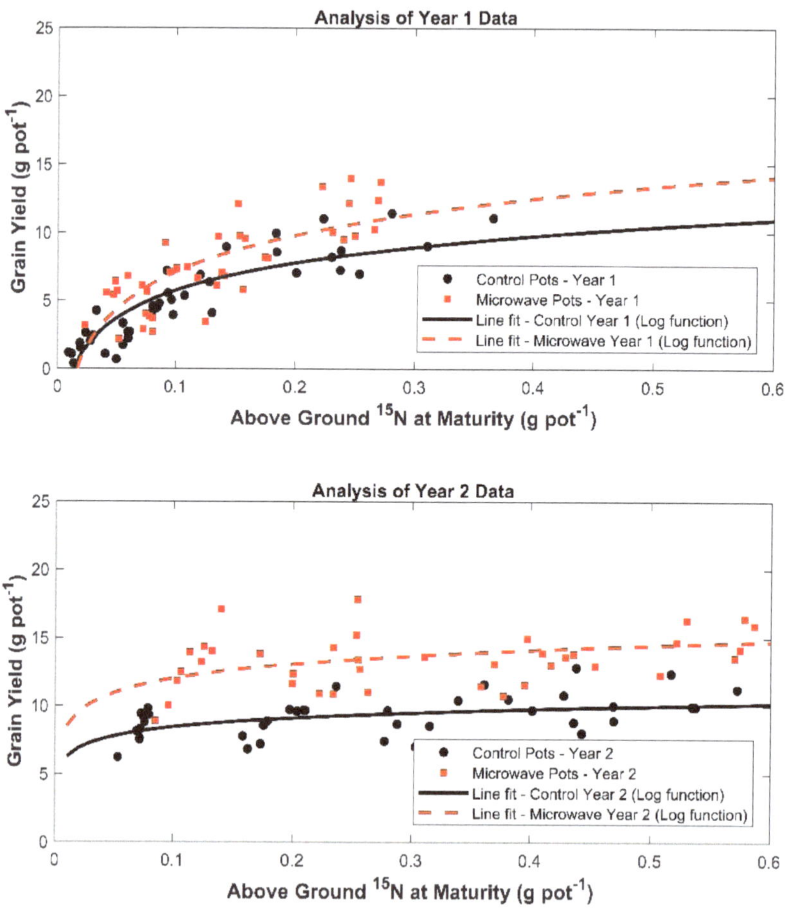


**Figure 11.11:** Grain yield as a function of above ground 15N at plant maturity.

## 11.4 Field Experiment – Rice at Dookie Campus

A field experiment was conducted at The University of Melbourne, Dookie College Campus in North East Victoria, Australia (-36.395 °S, 145.703 °E) to evaluate the effect of microwave (MW) irradiation of soil on weed emergence, plant grown and yield of rice. An area of 73.5 m$^2$ was excavated and manually levelled into a turkey-nest pond so the area could be flood irrigated to grow Rice. The experiment consists of two treatments: an untreated control ($T_0$) and microwave treated ($T_1$). The individual plots were 2.0 x 2.0 m, and were arranged with a 0.5 m untreated buffer zone between each plot, as shown in Figure 11.12.

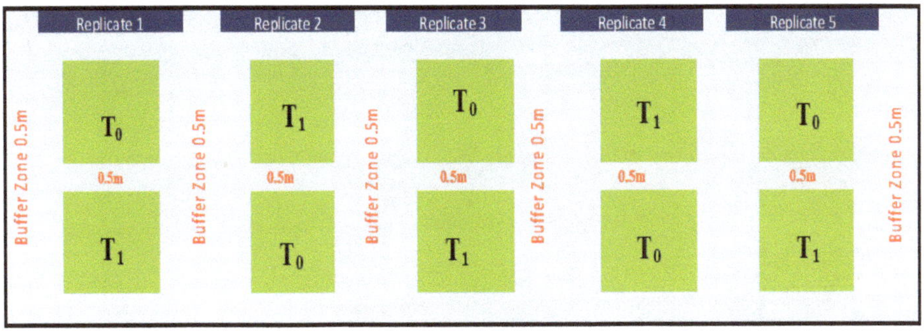


**Figure 11.12:** Experimental layout of the two microwave treatments: untreated control (T0) and MW treated (T1) of the Rice field experiment located at The University of Melbourne, Dookie College Campus in North East Victoria in 2015 – 16.

The soil of the microwave treated plots ($T_1$) was irradiated using two of the second microwave oven based prototype systems, described in Chapter 8. Two horn antennae with internal dimensions of 110 x 55 mm, which were separately attached to two domestic microwave ovens (EMS8586V; Sanyo; Tokyo, Japan) operating at 600 W with a frequency of 2.45 GHz. The applied microwave energy density in the treated plots, accounting for microwave energy reflections from the soil in the absence of using any wave-guide tuning, was approximately 730 J $cm^{-2}$ based on finite difference time domain (FD-TD) analyses. The soil was watered to field capacity prior to microwave treatment. Infrared thermal images, captured with an infrared camera (C2; FLIR Systems Inc; Wilsonville, Oregon, USA) immediately after treatment of the area under the horn, confirmed that the soil temperature exceeded 80°C (Figure 11.13).

The experimental plot was then flooded to a depth of 5 cm and the rice variety OPUS was broadcasted by hand at a seeding rate of 125 kg $ha^{-1}$. Phosphorus (80kg $ha^{-1}$), potassium (60kg $ha^{-1}$) and zinc (4 kg $ha^{-1}$) were applied to the entire experimental area at the time of sowing. Three split doses of nitrogen, equivalent to a total rate of 120 kg $ha^{-1}$, were applied during the growing season. Netting was fixed over the experimental plot to reduce bird scavenging during the early stages of development.

An infrared gas analyser (LI-6400XT; LI-COR Inc; Lincoln, Nebraska USA) was used to measure the physiological parameters of rice at maximum tillering. Number of tillers, fresh biomass and dry biomass were measured for a randomly selected 0.09 $m^2$ quadrat drawn from each of plot. Weed population was counted for each experimental plot. Yield estimates have been calculated based on sampling completed at about 85% physiological maturity, with final yield analysis expected to be completed in late April. Mean estimated grain yield was calculated for each experimental plot, based on measured number of fertile tillers per square metre, number of grains per spike, and weight of 100 grains (multiply sampled from each plot).

**Figure 11.13:** Prototype two horn antennae microwave system operated through two 2.45 GHz domestic microwave ovens (EMS8586V; Sanyo; Tokyo, Japan) used for soil irradiation in field conditions.

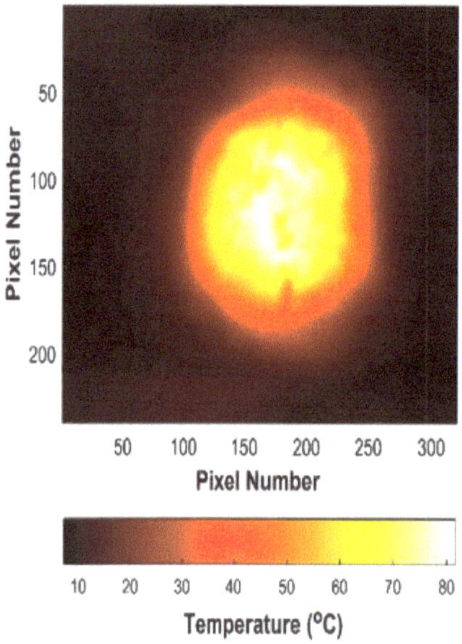


**Figure 11.14:** Infrared thermal image of the microwave heating pattern of the soil in field conditions for the rice experiment, irradiated through the prototype microwave system (Figure 11.13).

**Table 11.7:** Assessment of key crop growth parameters for rice crop experiment (Note: means with different superscripts on the same line are significantly different from one another) (Source: Khan, Brodie, & Dorin, 2017).

| | Treatment | | $LSD_{5\%}$ | % Change from control |
|---|---|---|---|---|
| | MW | C | | |
| Fresh Weight at panicle formation stage (g quadrate$^{-1}$) | 416.8$^{a}$ | 225.5$^{b}$ | 116.3 | 85% |
| Dry Weight at panicle formation stage (g quadrate$^{-1}$) | 91.3$^{a}$ | 50.8$^{b}$ | 26.1 | 80% |
| Tiller Density at panicle formation stage (number quadrate$^{-1}$) | 104.0$^{a}$ | 61.5$^{b}$ | 32.2 | 69% |
| Weed Density (number plot$^{-1}$) | 7.5$^{a}$ | 44.3$^{b}$ | 28.4 | -83% |
| Chlorophyll Content | 42.3 | 43.6 | 4.5 | -3% |
| Leaf Area Index | 4.0 | 2.6 | 2.0 | 56% |
| Yield (t ha$^{-1}$) | 9.0$^{a}$ | 6.7$^{b}$ | 1.7 | 34% |

Univariate analysis was undertaken with a statistical software package (16th Edition of Genstat; VSN International Limited; Oxford, UK) using general linear model analysis of variance (ANOVA) with MW treatment ($T_0$, $T_1$) as the fixed model terms; block was used as the random model term; and each measurement of interest (i.e. fresh weight) input as the response variable.

Microwave irradiation of the soil reduced weed emergence in the treated plots by 83%. Microwave treatment also enhanced many of the growth parameters including: the tiller density, which was visually evident (Figure 11.15 and Figure 11.16); crop fresh weight; dry biomass; and grain yield (Tab. 11.7); however, no significant difference was observed in chlorophyll content or leaf area index (Tab. 11.7).

## 11.5 Field Experiment – Rice at two Agroecological Zones of Australia

Two field trials were conducted from October, 2016 to April, 2017 using the same experimental layout as the first rice field experiment; however, these experiments were conducted at two different locations. The first location used the same site as before, at Dookie Campus of the University of Melbourne (36.395 °S, 145.703 °E). The second location Old Coree, Jerilderie, New South Vales (35.210 °S, 145.440 °E), which is the rice research farm, totally owned by Rice Research Australia Pty. Ltd. – SunRice™.

These experimental sites were treated using the trailer mounted microwave prototype, described in Chapter 8. In these two field trials microwave soil heating, did

**Figure 11.15:** Comparison of randomly sampled rice plants grown in the microwave treated plots (left) with rice plants grown in the control plots (right).

**Figure 11.16:** Comparison of microwave treated plot (left) with control plot (right).

not significantly reduce the weed establishment at both study location. However, the rice productivity increased in response to microwave application (Tab. 11.8 and Tab. 11.9). At Dookie, the microwave treatment gave no significant influence, in terms of final biomass (16.90 t $ha^{-1}$) and grain yield (3.88 t $ha^{-1}$), compare to control plots. At Jerilderie, microwave soil heating significantly ($P < 0.001$) augmented the rice biomass (19.80 t $ha^{-1}$) and grain yield (9.21 t $ha^{-1}$) compare to untreated control plots. The plant tiller density was greater in the microwave treated plots (Figure 11.17).

**Table 11.8:** Effect of pre-sowing microwave soil heating on growth and yield components of rice crop at Dookie Campus.

| Rice Parameters | Treatments | | LSD (p=0.05) | P-Value | Percentage Change |
|---|---|---|---|---|---|
| | Microwave Treated | Untreated Control | | | |
| Dry Biomass Weight ($t\ ha^{-1}$) | $16.90^{a}$ | $14.00^{a}$ | 4.5 | 0.19 | 20.71% |
| Grain Yield ($t\ ha^{-1}$) | $3.9^{a}$ | $2.4^{a}$ | 1.5 | 0.06 | 61.67% |
| Harvest Index | 22.26 a | 17.12 a | 6.41 | 0.11 | 6.10% |

**Table 11.9:** Effect of pre-sowing microwave soil heating on growth and yield components of rice crop at Old Coree, Jerilderie, New South Wales.

| Rice Parameters | Treatments | | LSD (p=0.05) | P-Value | Percentage Change |
|---|---|---|---|---|---|
| | Microwave Treated | Untreated Control | | | |
| Dry Biomass Weight (t ha–1) | 19.80 a | 17.05 b | 0.57 | <0.001 | 16.10% |
| Grain Yield (t ha–1) | 9.21 a | 7.63 b | 0.65 | <0.001 | 20.70% |
| Harvest Index | 46.77 a | 44.59 a | 4.29 | 0.26 | 6.10% |

## 11.6 Field Experiment – Tomato

A field experiment was established in a commercial tomato crop in Corop, Victoria (-36.4577 °S, 144.7965 °E) to compare the effects of standard practice (Soil Fumigation), microwave treatment and no treatment (control) on weed emergence, plant growth and yield. The tomatoes are grown in 1.0 x 3.4 m raised beds, with a 1.0 m trench between each of the raised beds. Standard practice for the crop involves treating the soil with the soil fumigant Metham Sodium prior to planting. An area comprising five 1.0 m x 3.4 m raised beds was not fumigated. The non-fumigated area was split into 10 experimental plots (1.0 x 1.7 m) and the treatments: an untreated control ($T_0$) and microwave treated ($T_1$), were arranged in a randomised complete block design (RCBD). The microwave treated plots ($T_1$) were treated with a 546 J $cm^{-2}$ of microwave energy using horn antennae prior to planting.

**Figure 11.17:** Comparison of early growth establishment of rice crop. Plants on left collected from microwave treated plot and plants on right collected from untreated control plot. (Left image taken from Dookie Trial Site and right image taken from Old Coree, Jerilderie site.).

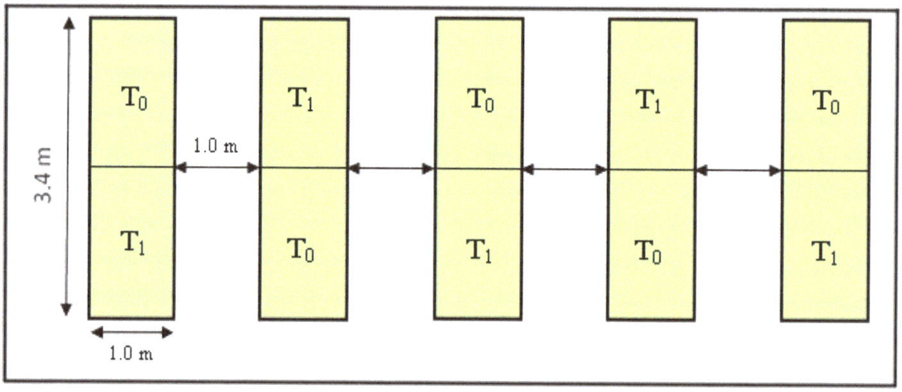


**Figure 11.18:** Experimental layout of the two microwave treatments: untreated control (T0) and MW treated (T1) of the Tomato crop located at Corop in North East Victoria in 2015 – 16.

Weed emergence per plot, fruit number per plant, and flower number per plant were monitored regularly throughout the growing season. Final biomass and fruit yields were assessed at harvest in March, 2016.

Microwave treatment significantly reduced weed emergence (83%) compared to control and chemically treated plots, however no significant difference was observed between the control and chemically treated plots (Tab. 11.10). Mean fruit yield in the microwave treated plots was 40% higher than in the control plots and 37% higher than in the fumigated soil plots (Tab. 11.10). There was a significantly higher mean

**Table 11.10:** Mean crop parameters for untreated control (C), microwave treated (MW) and standard practice – fumigated (F) plots in the commercial tomato experiment located at Corop in Victoria, Australia. Dissimilar lower case letters denote significant difference at 5%.

| | Treatment | | | $LSD_{5\%}$ | % Change from control | % Change from chemical |
|---|---|---|---|---|---|---|
| | MW | C | F | | | |
| Mean Number of Fruit per plant | $187.30^{a}$ | $106.40^{b}$ | $149.25^{ab}$ | 43.93 | 76% | 25% |
| Mean Number of Weeds per plot | $0.50^{a}$ | $3.00^{b}$ | $2.50^{b}$ | 1.71 | -83% | -80% |
| Mean Fresh Tomato Yield (kg plot$^{-1}$) | $30.45^{a}$ | $21.79^{b}$ | $22.17^{ab}$ | 8.38 | 40% | 37% |
| Mean Biomass (kg plot$^{-1}$) | $35.94^{a}$ | $25.32^{b}$ | $25.22^{b}$ | 9.75 | 42% | 43% |

number of fruit per plant and biomass for microwave treated plots compared to the Control plots; however, there was no significant difference between Fumigated soil and microwave or Control plots (Tab. 11.10).

## 11.7 Field Experiment – Wheat (Dookie)

A field experiment was conducted at The University of Melbourne, Dookie College Campus in North East Victoria, Australia (-36.395 °S, 145.703 °E) to evaluate the effect of microwave irradiation of soil on weed emergence, plant growth and yield of wheat. The same experimental layout and microwave treatment protocol was applied to the soil as was used in the rice trial conducted at Dookie in Experiment 4. Soil at the experimental site is a sandy loam and classified as a Currawa Loam (Downes, 1949) or a Yellow Subnatric-Dystropic Sodosols (Isbell, 2002). The soil properties (0 to 15 cm) were as follow: bulk density (1.6 gm cm$^{-3}$), Nitrate-Nitrogen 49 mg kg$^{-1}$, Phosphorus $_{(Colwell)}$ 31 mg kg$^{-1}$, $pH_{(H2O)}$ 5.6, Electrical Conductivity (EC) 0.28 dS m$^{-1}$, and Cation Exchange Capacity (CEC) 10.20 meq 100 g$^{-1}$.

Historically, the same paddock has been used extensively to grow crops in a monocropping pattern with wheat-canola rotations. Gregory winter Wheat was sown on the 6$^{th}$ of June, 2016, using a cone seeder. The Gregory wheat variety was sown using a small plot seeder, with row to row spacing of about 25 cm at the seed rate of 60 kg ha$^{-1}$. Planting occurred in a last week of May, 2016. All the agronomic practices, except herbicides application, were followed according to Goulburn Valley Wheat Belt region of North Victoria, Australia. Diammonia Phosphate (80 kg ha$^{-1}$) and potassium

(60 kg $ha^{-1}$) were applied to the entire experimental area at the time of sowing. Three split doses of nitrogen applications, equivalent to a total rate of 120 kg $ha^{-1}$, using urea, were applied during the growing season. The number of tillers, fresh biomass and dry biomass were measured for a randomly selected 0.25 $m^2$ quadrats, drawn from each plot, for early growth assessment in response to MW soil heating and weights were converted to tonnes per hectare. At physiological maturity, the crop was manually harvested from whole plots for yield assessment. After harvesting, the crop biomass was dried in an air circulation oven (Nabertherm; TR1050 27124; Germany) for 24 h at 65°C and final crop yield was converted into tonnes per hectare. Weed density and biomass accumulation was counted from whole experimental plot areas (2.6 $m^2$) at maximum tillering stage and dry biomass of weeds were assessed at crop harvest. Harvest index was calculated by using equation (11.4).

$$Harvest\ Index = \frac{Grain\ Yield}{Biological\ Yield} \times 100 \tag{11.4}$$

Microwave soil treatment significantly reduced weed density and weed biomass compared with the untreated control. Those weeds, which emerged in the microwave treated plots, were very small compared to those in the untreated control plots. Tiller density and crop biomass were both significantly increased in the microwave treated plots, compared with the control plots (Tab. 11.11). The application of microwave treatment to the soil for pre-emergence weed control significantly increased the dryland wheat productivity.

Early crop weight gain is considered to be a crop-stability trait in dryland agricultural systems. The 50.9% gain in fresh weight and 42.4% gain in dry weight were acquired from the microwave treated plots. Less weed-crop competition for vital resources may be the possible cause of good crop growth in the microwave treated scenario, because weeds have a direct impact on wheat production. Relevant to crop weight gain, the current findings are favourably supported by Gibson, Fox, and Deacon (1988). They reported that microwave soil treatment markedly enhanced the shoot weight of birch (*Betula pendula*) seedlings; the maximum shoot weight (84 mg) was attained through exposure to microwave (600 W; 120 s) compared to the untreated control soil (25 mg). The wheat growth and weed suppression in response to microwave treated and untreated plots are shown in Figure 11.19.

Tiller density is the main contributor of final grain yield. The maximum number of tillers was developed in the microwave treated plot (387 $m^{-2}$) compared to the untreated control plots (268 $m^{-2}$). In the present investigation, the final incremental increase of 33.1% in dry biomass production and 39.2% in grain yield was attained through MW energy application into the soil for weed management.

**Table 11.11:** Effect of pre-sowing microwave soil treatment on growth and yield of wheat crop under field conditions.

| Wheat Parameters | Treatments Microwave Treated | Untreated Control | LSD (p=0.05) | Percentage Change |
|---|---|---|---|---|
| Number of Tillers (m$^{-2}$)[i] | 387 $^{a*}$ | 268 $^{b}$ | 62 | 44.40% |
| Fresh Biomass Weight (t ha$^{-1}$)[i] | 30.8 $^{a}$ | 20.4 $^{a}$ | 10.4 | 50.90% |
| Dry Biomass Weight (t ha$^{-1}$)[i] | 4.7 $^{a}$ | 3.3 $^{b}$ | 1.5 | 42.40% |
| Dry Biomass Weight (t ha$^{-1}$)[ii] | 19.7 $^{a}$ | 14.8 $^{b}$ | 4.8 | 33.10% |
| Grain Yield (t ha$^{-1}$)[ii] | 7.8 $^{a}$ | 5.6 $^{b}$ | 2.3 | 39.20% |
| 1000 Grain Weight (gm)[ii] | 45.9 $^{a}$ | 42.8 $^{b}$ | 3.2 | 7.20% |
| Harvest Index | 39.9 $^{a}$ | 37.6 $^{a}$ | 6.1 | 6.10% |

[i] Data collection at maximum tillers establishment.
[ii] Data collection at crop harvesting.

**Figure 11.19:** Comparison of weed seedling establishment in microwave treated plot (left) and untreated control plot (right) in dryland wheat at Dookie.

## 11.8 General Discussion & Conclusion

Microwave pre-treatment of the soil, prior to crop planting has been shown to reduce weed emergence and enhance plant vigour and increase final yield potential in glass house and field conditions. The glass house microwave-nitrogen interaction experiments reveal that yield response to microwave soil treatment is dependent on the application rate of nitrogen to the crop. This suggests that microwave treatment of the soil, prior to crop sowing, changes the soil nitrogen availability; hence, as more chemical nitrogen is applied, the plants rely less on nitrogen that has been made available by the microwave soil treatment process (Figure 11.6). Despite this, there is still a residual 30 % increase in crop yield potential, due to microwave soil treatment, as the application of chemical nitrogen becomes larger (Note: the curve in Figure

11.6 asymptotes to 0.3, not zero). It is also evident that there is an ongoing benefit of microwave soil treatment beyond a single growing season.

Microwave treatment of soil significantly increased the grain yield, where no nitrogen application was done. This contribution in grain yield may be acquired from indigenous soil nitrogen. The direct and indirect effect of applied N ($^{15}$N-lablled) in microwave treated soil contributed to a marginal benefit of grain yield. The $Y_0/N_r$ in 2015 from microwave treated pots (5.68 gm pot$^{-1}$) was 171.8% higher compare to untreated control soil (2.06 gm pot$^{-1}$). This suggests that the availability and acquisition of indigenous nitrogen for grain production.

The field trials have all demonstrated significant reductions in weed emergence during the cropping period due to pre-treatment of the soil with microwave energy. Significant yield increases were also achieved from microwave pre-treatment in the field trials. It is interesting to note that microwave pre-treatment of the soil in the commercial tomato crop site provided better weed control and fruit yield than the standard practice of soil fumigation adopted in the remainder of the crop.

Therefore, in addition to weed suppression, a few previous studies have reported the supplementary effect of microwave energy on soil nutrient dynamics; Yang, Skogley, and Schaff (1990) tested the nutrient extractability effect of microwave on soil. When fresh soil was exposed to microwave energy a dramatic increase in the $NH_4^+$-N concentration was observed for an extended treatment of 120 sec. They concluded that this effect was partially from non-microbial processes, either from site exchange or from fixed position in inorganic collides (clay minerals). Hur, Park, Kim, and Kim (2013) demonstrated that microwave irradiation of soil can enhance the binding efficiency of hydrophobic organic containments with soil organic matter. They irradiated 5 g samples of soil in plastic tubes in aerobic and anaerobic conditions with activated C for 600 s in a lab-scale MW oven (2.45 GHz) operated at 700W. They pointed out that microwave irradiation significantly alters the physical and chemical properties of soil organic matter and increased its humification. In another study, Kim and Kim (2013) studied the influences of microwave irradiation on the soil organic matter properties. They reported that thermal cracking induced by irradiation potentially alters the molecular composition (C, H, O and N), chemical structure and humification of soil organic matter. Based on these previous findings, we assumed that thermal denaturation of recalcitrant humic substance induced by microwave irradiation may increase the concentrations of free amino acids for succeeding turnover to $CO_2$ and ammonia pool $NH_4^+$, which might have substantially increased wheat productivity in the present investigation. However, in addition to weed control, future research should also elucidate the effect of MW soil treatment on soil nutrient dynamics and associated microbial activity.

In conclusion, microwave pre-treatment of the soil, prior to crop planting, significantly reduces weed emergence and significantly increases crop yield potential (Tab. 11.12).

**Table 11.12:** Summary of crop yield responses to microwave treatment.

| Microwave Treatment (J cm$^{-2}$) | Control | Hand Weeded | 80 | 160 | 320 | LSD (P = 0.05) | Change from Hand Weeded/ Control |
|---|---|---|---|---|---|---|---|
| *Pot Trials* | | | | | | | |
| Canola Pod Yield (g pot$^{-1}$) | 0.27$^{a}$ | 0.56$^{a}$ | 0.36$^{a}$ | 1.25$^{b}$ | 1.95$^{c}$ | 0.55 | **250%** |
| Wheat Grain Yield (g pot$^{-1}$) | 0.66$^{a}$ | 0.67$^{a}$ | 0.68$^{a}$ | 0.75$^{a}$ | 1.25$^{b}$ | 0.3 | **87%** |
| Rice Grain Yield (g pot$^{-1}$) | 40.0$^{a}$ | 41.3$^{a}$ | 43.3$^{a}$ | 59.0$^{ab}$ | 64.0$^{b}$ | 18.9 | **55%** |
| Maize (g pot$^{-1}$) | 5.3$^{a}$ | 6.6$^{a}$ | — | 10.3$^{ab}$ | 12.8$^{b}$ | 4.8 | **92%** |
| *Field Trials* | | | | | | | |
| Rice (t ha$^{-1}$) – Dookie Year 1 (2015/2016) | 7.5$^{a}$ | — | — | — | 10.1$^{b}$ | 2 | **35%** |
| Rice (t ha$^{-1}$) – Dookie Year 2 – (2016/2017) - cold affected | 2.1$^{a}$ | — | — | — | 3.9$^{b}$ | 1.3 | **84%** |
| Rice (t ha$^{-1}$) – Old Coree – (2016/2017) | 7.7$^{a}$ | — | — | — | 9.1$^{b}$ | 1.2 | **19%** |
| Wheat (t ha$^{-1}$) | 5.7$^{a}$ | — | — | — | 7.8$^{b}$ | 1.4 | **39%** |
| Tomato (t ha$^{-1}$) | 64.1$^{a}$ | 65.2$^{a}$ | — | — | 89.6$^{b}$ | 24.7 | **37%** |

## 11.9 References


Brodie, G., Bootes, N., & Reid, G. (2015). Invited Paper. Plant growth and yield of wheat and canola in microwave treated soil. Paper presented at the IMPI's 49th Microwave Power Symposium, San Diego, California, USA.

Brodie, G., Grixti, M., Hollins, E., Cooper, A., Li, T., & Cole, M. (2015). Assessing the Impact of Microwave Treatment on Soil Microbial Populations. Global Journal of Agricultural Innovation, Research & Development, 2(1), 25-32.

Brodie, G., Hamilton, S., & Woodworth, J. (2007). An assessment of microwave soil pasteurization for killing seeds and weeds. Plant Protection Quarterly, 22(4), 143-149.

Brodie, G., Hollins, E., & Woods-Casey, J. (2014). Recent Work on Microwave-based Weed Control for Cropping Systems. Paper presented at the 19th Australasian Weed Conference, Hobart, Australia.

Brodie, G., Ryan, C., & Lancaster, C. (2012). Microwave technologies as part of an integrated weed management strategy: A Review. International Journal of Agronomy, 2012, 1-14.

Carter, C. A., Chalfant, J. A., Goodhue, R. E., Han, F. M., & DeSantis, M. (2005). The Methyl Bromide Ban: Economic Impacts on the California Strawberry Industry. Review of Agricultural Economics, 27(2), 181-197. doi:http://www.blackwellpublishing.com/journal.asp?ref=1058-7195

Diprose, M. F., Benson, F. A., & Willis, A. J. (1984). The Effect of Externally Applied Electrostatic Fields, Microwave Radiation and Electric Currents on Plants and Other Organisms, with Special Reference to Weed Control. Botanical Review, 50(2), 171-223.

Downes, R. G. (1949). A soil, land-use and erosion survey of parts of the counties of Moira and Delatite, Victoria. Melbourne: CSIRO.

Fennimore, S. A., Haar, M. J., & Ajwa, H. A. (2003). Weed Control in Strawberry Provided by Shank- and Drip-applied Methyl Bromide Alternative Fumigants. HortScience, 38(1), 55-61.

Francis, C. A., Roberts, K. J., Beman, J. M., Santoro, A. E., & Oakley, B. B. (2005). Ubiquity and diversity of ammonia-oxidizing archaea in water columns and sediments of the ocean. Proceedings of the National Academy of Sciences of the United States of America, 102(41), 14683-14688.

Gibson, F., Fox, F. M., & Deacon, J. W. (1988). Effects of microwave treatment of soil on growth of birch (Betula pendula) seedlings and infection of them by ectomycorrhizal fungi. New Phytology, 108(2), 189-204. doi:https://dx.doi.org/10.1111/j.1469-8137.1988.tb03696.x

Hur, J., Park, S.-W., Kim, M. C., & Kim, H. S. (2013). Enhanced binding of hydrophobic organic contaminants by microwave assisted humification of soil organic matter. Chemosphere, 93(11), 2704-2710.

Isbell, R. F. (2002). The Australian soil classification. Melbourne: CSIRO.

Khan, M. J., Brodie, G., & Dorin, G. (2017). The Effect of Microwave Soil Treatment on Rice Production under Field Conditions. Transactions of the American Society of Agricultural and Biological Engineers, 60(2), 517-525. doi:10.13031/trans.12125

Khan, M. J., Brodie, G., & Gupta, D. (2016). Effect of Microwave (2.45 GHz) Treatment of Soil on Yield Components of Wheat (Triticum aestivum L.). Journal of Microwave Power and Electromagnetic Energy, 50(3), 191–200.

Kim, M. C., & Kim, H. S. (2013). Artificial and enhanced humification of soil organic matter using microwave irradiation. Environmental Science and Pollution Research, 20(4), 2362-2371.

Rotthauwe, J. H., Witzel, K. P., & Liesack, W. (1997). The ammonia monooxygenase structural gene amoA as a functional marker: molecular fine-scale analysis of natural ammonia-oxidizing populations. Applied and Environmental Microbiology, 63(12), 4704-4712.

Salamanca-Jimenez, A., Doane, T. A., & Horwath, W. R. (2017). Nitrogen use efficiency of coffee at the vegetative stage as influenced by fertilizer application method. Frontiers in Plant Science, 8.

Sances, F. V., & Ingham, E. R. (1997). Conventional and organic alternatives to methyl bromide on california strawberries. Compost Science and Utilization, 5(2), 23-38.

Sydorovych, O., Safley, C. D., Ferguson, L. M., Poling, E. B., Fernandez, G. E., Brannen, P. A., et al. (2006). Economic evaluation of methyl bromide alternatives for the production of strawberries in the southeastern United States. HortTechnology, 16(1), 118 -128.

Yang, J. E., Skogley, E. O., & Schaff, B. E. (1990). Microwave Radiation and Incubation Effects on Resin-Extractable Nutrients: I. Nitrate, Ammonium, and Sulfur. Soil Science Society of America Journal, 54(6), 1639-1645.

## 11.10 Appendices

**Table 11.13:** Description of the Gowangardie Soil (Downes, 1949), which is typical to the experimental site.

| **Gowangardie Loam** | | |
|---|---|---|
| **Horizon** | **Depth** | **Description** |
| $A_1$ | 0 – 10.16 cm (0 – 4 Inches) | Brownish-grey loam with some pieces of iron-impregnated parent rock |
| $A_2$ | 10.16 – 25.4 cm (4 – 9 Inches) | Light grey-brown loam to sandy loam with some pieces of iron-impregnated parent rock. |
| $B_1$ | 10.16 – 68.58 cm (9 – 27 Inches) | Brown to red-brown with slight grayish mottling, heavy clay. Small nutty structure when dry, but sticky, plastic when wet. |
| $B_2C$ | 68.58 – 91.44 cm (27 – 36 Inches) | Brown to red-brown clay with purple and yellow pieces of decomposing parent rock. |

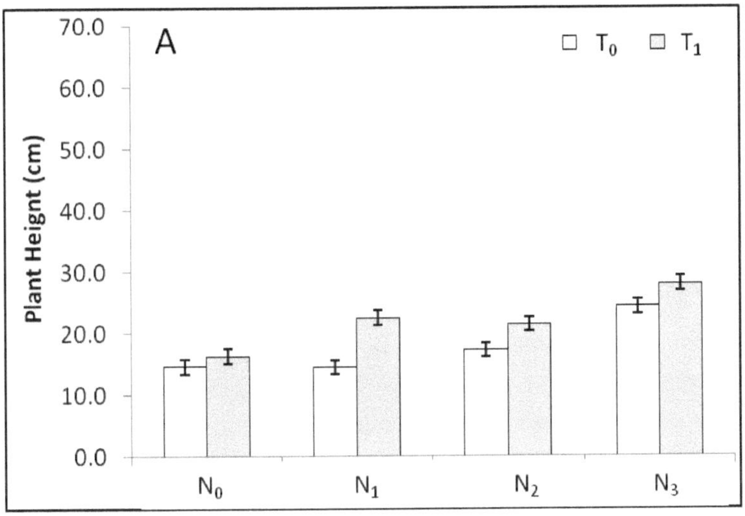


**Figure 11.20:** Plant height (cm) of Wheat grown in a glasshouse as function microwave (MW) treated soil ($T_0$ = untreated (control); $T_1$ = 120 sec 2.45GHz MW treatment under different rates of N application ($N_0$ = 0 mg pot$^{-1}$; $N_1$= 22.5 mg pot$^{-1}$; $N_2$ = 45.0 mg pot$^{-1}$; $N_3$= 67.5 mg pot$^{-1}$) measured at (A) 20 days after sowing (DAW); (B) 40 DAW; and (C) 60 DAW. Error bars indicate LSD at 5

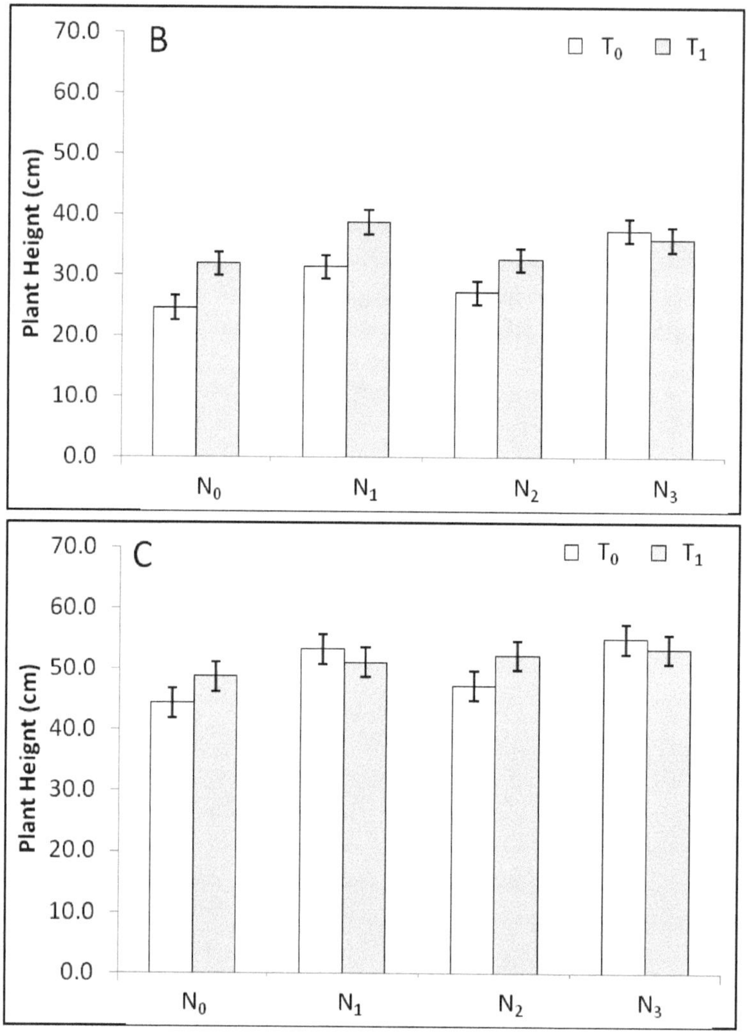


Continued **Figure 11.20:** Plant height (cm) of Wheat grown in a glasshouse as function microwave (MW) treated soil ($T_0$ = untreated (control); $T_1$ = 120 sec 2.45GHz MW treatment under different rates of N application ($N_0$ = 0 mg pot$^{-1}$; $N_1$= 22.5 mg pot$^{-1}$; $N_2$ = 45.0 mg pot$^{-1}$; $N_3$= 67.5 mg pot$^{-1}$) measured at (A) 20 days after sowing (DAW); (B) 40 DAW; and (C) 60 DAW. Error bars indicate LSD at 5

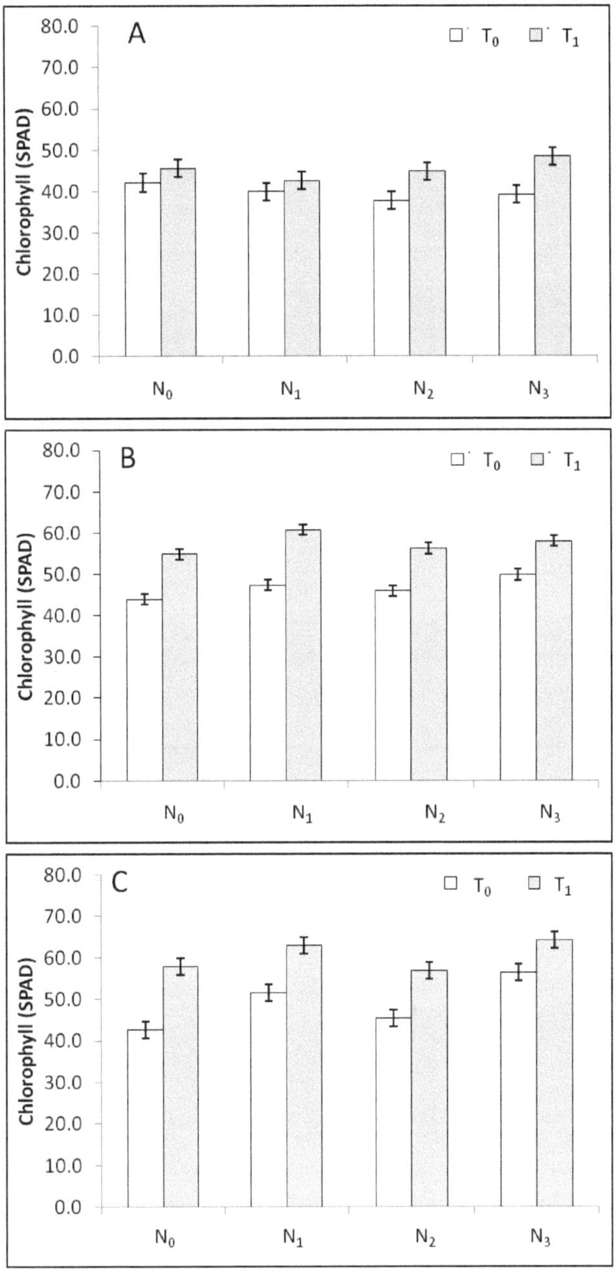


**Figure 11.21:** Chlorophyll content (SPAD) of Wheat grown in a glasshouse as function microwave (MW) treated soil ($T_0$ = untreated (control); $T_1$ = 120 sec 2.45GHz MW treatment under different rates of N application ($N_0$ = 0 mg pot$^{-1}$; $N_1$= 22.5 mg pot$^{-1}$; $N_2$ = 45.0 mg pot$^{-1}$; $N_3$ = 67.5 mg pot$^{-1}$) measured at (A) 20 days after sowing (DAW); (B) 40 DAW; and (C) 60 DAW. Error bars indicate LSD at 5%.

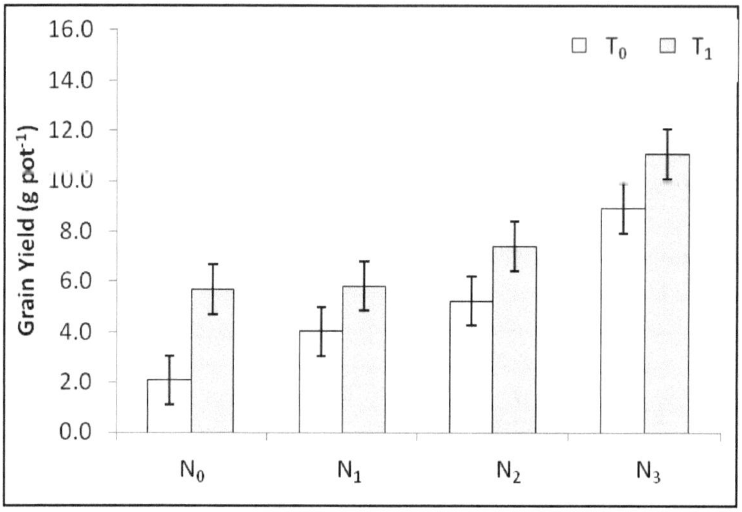


**Figure 11.22:** Grain yield per pot (g pot$^{-1}$) of Wheat grown in a glasshouse as a function microwave (MW) treated soil ($T_0$ = untreated (control); $T_1$ = 120 sec 2.45GHz MW treatment under different rates of N application ($N_0$ = 0 mg pot$^{-1}$; $N_1$= 22.5 mg pot$^{-1}$; $N_2$ = 45.0 mg pot$^{-1}$; $N_3$= 67.5 mg pot$^{-1}$) measured at harvest. Error bars indicate LSD at 5%.

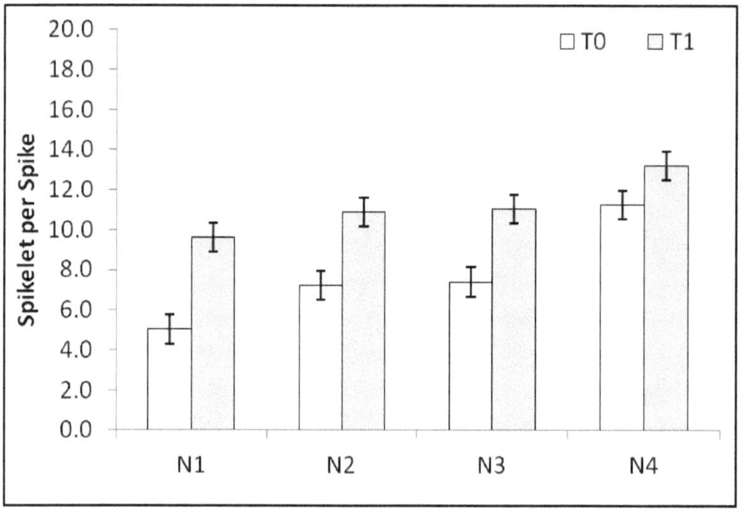


**Figure 11.23:** Spikelets per spike of Wheat grown in a glasshouse as a function microwave (MW) treated soil ($T_0$ = untreated (control); $T_1$ = 120 sec 2.45GHz MW treatment under different rates of N application ($N_0$ = 0 mg pot$^{-1}$; $N_1$= 22.5 mg pot$^{-1}$; $N_2$ = 45.0 mg pot$^{-1}$; $N_3$= 67.5 mg pot$^{-1}$) measured at harvest. Error bars indicate LSD at 5%.

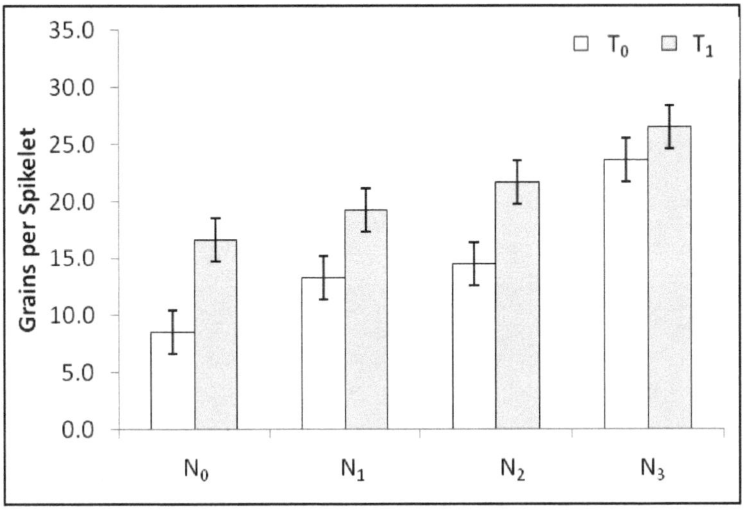


**Figure 11.24:** Grains per spikelet of Wheat grown in a glasshouse as a function microwave (MW) treated soil ($T_0$ = untreated (control); $T_1$ = 120 sec 2.45GHz MW treatment under different rates of N application ($N_0$ = 0 mg pot$^{-1}$; $N_1$= 22.5 mg pot$^{-1}$; $N_2$ = 45.0 mg pot$^{-1}$; $N_3$ = 67.5 mg pot$^{-1}$) measured at harvest. Error bars indicate LSD at 5%.

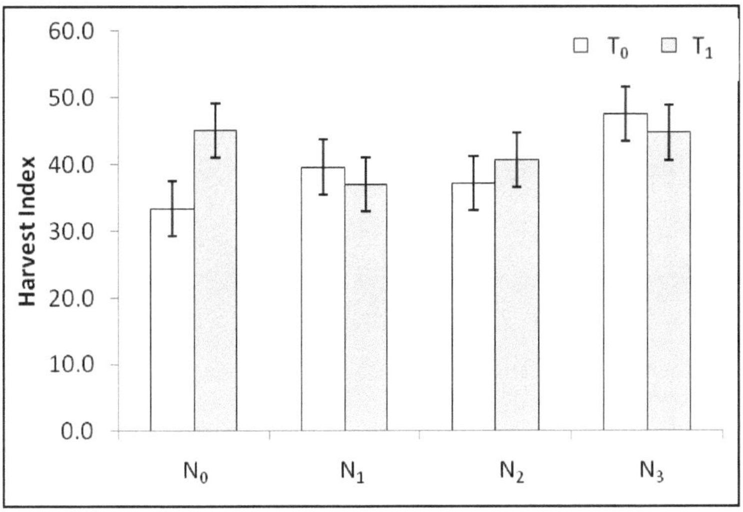


**Figure 11.25:** Harvest Index (grain yield/biomass yield*100) of Wheat grown in a glasshouse as a function microwave (MW) treated soil ($T_0$ = untreated (control); $T_1$ = 120 sec 2.45GHz MW treatment under different rates of N application ($N_0$ = 0 mg pot$^{-1}$; $N_1$= 22.5 mg pot$^{-1}$; $N_2$ = 45.0 mg pot$^{-1}$; $N_3$ = 67.5 mg pot$^{-1}$) measured at harvest. Error bars indicate LSD at 5%.

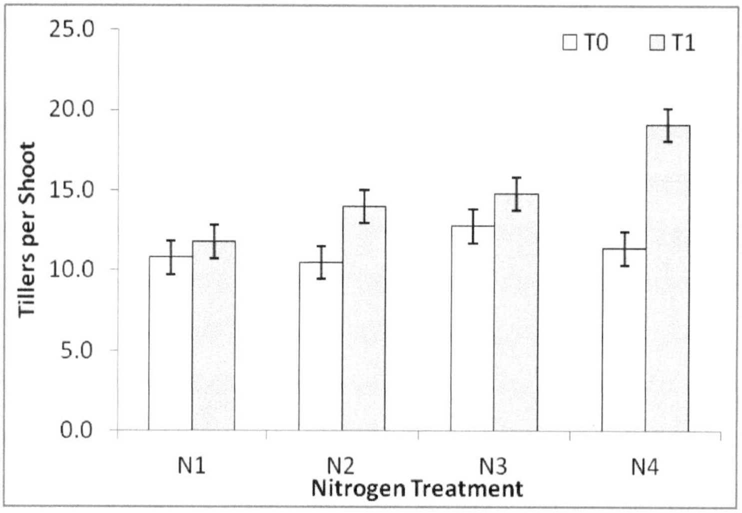


**Figure 11.26:** Tillers per shoot of Wheat grown in a glasshouse as a function microwave (MW) treated soil ($T_0$ = untreated (control); $T_1$ = 120 sec 2.45GHz MW treatment under different rates of N application ($N_0$ = 0 mg pot$^{-1}$; $N_1$= 22.5 mg pot$^{-1}$; $N_2$ = 45.0 mg pot$^{-1}$; $N_3$= 67.5 mg pot$^{-1}$) measured at harvest. Error bars indicate LSD at 5%.

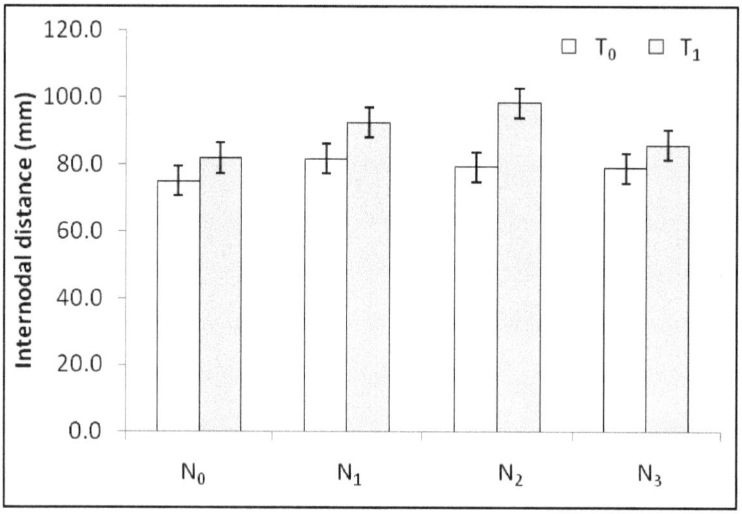


**Figure 11.27:** Internodal distance (mm) of Wheat grown in a glasshouse as a function microwave (MW) treated soil ($T_0$ = untreated (control); $T_1$ = 120 sec 2.45GHz MW treatment under different rates of N application ($N_0$ = 0 mg pot$^{-1}$; $N_1$= 22.5 mg pot$^{-1}$; $N_2$ = 45.0 mg pot$^{-1}$; $N_3$= 67.5 mg pot$^{-1}$) measured at harvest. Error bars indicate LSD at 5%.

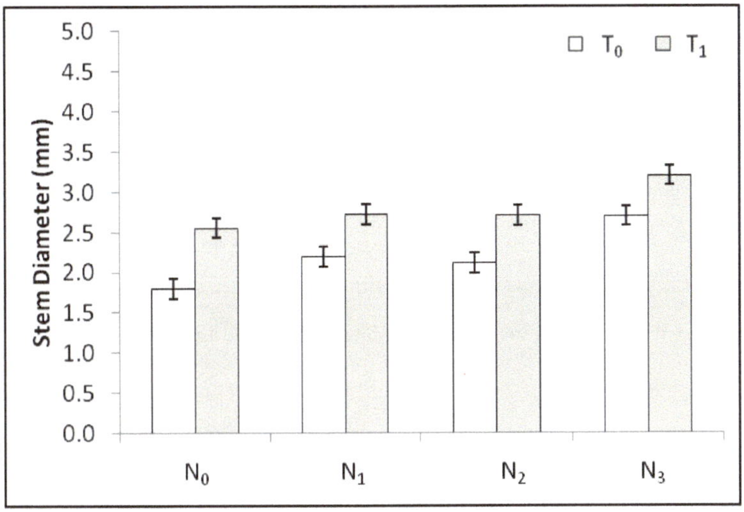


**Figure 11.28:** Stem diameter (mm) of Wheat grown in a glasshouse as a function microwave (MW) treated soil ($T_0$ = untreated (control); $T_1$ = 120 sec 2.45GHz MW treatment under different rates of N application ($N_0$ = 0 mg pot$^{-1}$; $N_1$= 22.5 mg pot$^{-1}$; $N_2$ = 45.0 mg pot$^{-1}$; $N_3$ = 67.5 mg pot$^{-1}$) measured at harvest. Error bars indicate LSD at 5%.

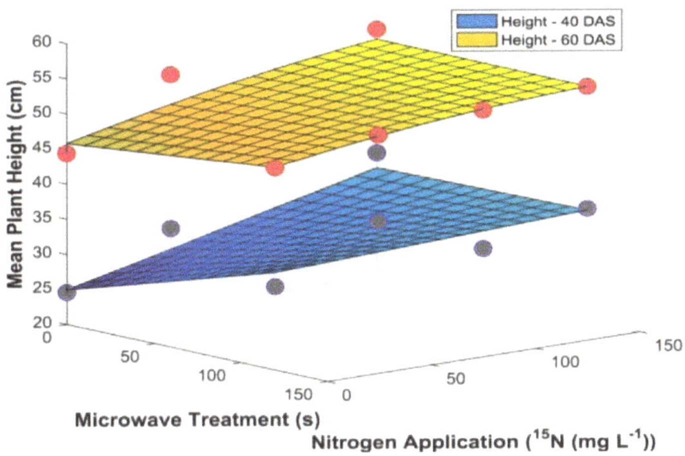


**Figure 11.29:** Effect of Microwave and various level of Nitrogen on Plant Height of Wheat (40 & 60 DAS).

**Table 11.14:** Mean plant height (cm).

| Microwave Treatment ($J\ cm^{-3}$) | Nitrogen Treatment ($mg\ L^{-1}$ of Solution) | | | |
|---|---|---|---|---|
| | **0** | **50** | **100** | **150** |
| 0 | $43.06^{a}$ | $46.41^{ab}$ | $46.52^{b}$ | $41.76^{a}$ |
| 60 | $50.92^{c}$ | $49.09^{bc}$ | $48.14^{bc}$ | $49.84^{bc}$ |
| LSD (P = 0.05) | | | | 3.35 |

Note: Means with different superscripts are significantly different from one another ($P < 0.05$)

**Table 11.15:** Mean Chlorophyll content (SPAD).

| Microwave Treatment ($J\ cm^{-3}$) | Nitrogen Treatment ($mg\ L^{-1}$ of Solution) | | | |
|---|---|---|---|---|
| | **0** | **50** | **100** | **150** |
| 0 | $42.74^{b}$ | $43.97^{b}$ | $44.25^{b}$ | $39.19^{a}$ |
| 60 | $44.42^{b}$ | $44.05^{b}$ | $44.4^{b}$ | $43.28^{b}$ |
| LSD (P = 0.05) | | | | 1.9 |

Note: Means with different superscripts are significantly different from one another ($P < 0.05$)

**Table 11.16:** Mean plant biomass yield from four sampled plants per pot ($g\ pot^{-1}$).

| Microwave Treatment ($J\ cm^{-3}$) | Nitrogen Treatment ($mg\ L^{-1}$ of Solution) | | | |
|---|---|---|---|---|
| | **0** | **50** | **100** | **150** |
| 0 | $0.34^{a}$ | $0.49^{ab}$ | $0.44^{ab}$ | $0.45^{ab}$ |
| 60 | $0.74^{bc}$ | $0.76^{c}$ | $0.59^{b}$ | $0.69^{bc}$ |
| LSD (P = 0.05) | | | | 0.16 |

Note: Means with different superscripts are significantly different from one another ($P < 0.05$)

**Table 11.17:** Mean flag leaf length (cm).

| Microwave Treatment ($J\ cm^{-3}$) | Nitrogen Treatment (mg $L^{-1}$ of Solution) | | | |
|---|---|---|---|---|
| | 0 | 50 | 100 | 150 |
| 0 | $21.55^{b}$ | $18.35^{b}$ | $19.31^{b}$ | $14.5^{a}$ |
| 60 | $23.48^{c}$ | $18.88^{b}$ | $21.66^{bc}$ | $20.43^{bc}$ |
| LSD (P = 0.05) | | | | 3.36 |

Note: Means with different superscripts are significantly different from one another (P < 0.05)

**Table 11.18:** Mean dry biomass at harvest (g $pot^{-1}$).

| Microwave Treatment ($J\ cm^{-3}$) | Nitrogen Treatment (mg $L^{-1}$ of Solution) | | | |
|---|---|---|---|---|
| | 0 | 50 | 100 | 150 |
| 0 | $35.46^{bc}$ | $35.75^{bc}$ | $33.88^{b}$ | $23.68^{a}$ |
| 60 | $51.05^{a}$ | $46.96^{c}$ | $42.12^{c}$ | $40.93^{c}$ |
| LSD (P = 0.05) - Microwave | | | | 6.38 |

Note: Means with different superscripts are significantly different from one another (P < 0.05)

**Table 11.19:** Mean grain yield at harvest (g $pot^{-1}$).

| Microwave Treatment ($J\ cm^{-3}$) | Nitrogen Treatment (mg $L^{-1}$ of Solution) | | | |
|---|---|---|---|---|
| | 0 | 50 | 100 | 150 |
| 0 | $18.33^{bc}$ | $18.56^{bc}$ | $16.26^{b}$ | $10.75^{a}$ |
| 60 | $26.73^{cd}$ | $24.99^{c}$ | $22.09^{c}$ | $20.34^{bc}$ |
| LSD (P = 0.05) - Microwave | | | | 4.29 |

Note: Means with different superscripts are significantly different from one another (P < 0.05)

**Table 11.20:** Mean plant height (cm).

| Microwave Treatment (J cm$^{-3}$) | Nitrogen Treatment (mg L$^{-1}$ of Solution) | | | |
|---|---|---|---|---|
| | 0 | 50 | 100 | 150 |
| 0 | 32.65$^{a}$ | 33.07$^{a}$ | 34.95$^{b}$ | 32.41$^{a}$ |
| 60 | 37.07$^{c}$ | 37.5$^{c}$ | 37.59$^{c}$ | 36.61$^{bc}$ |
| LSD (P = 0.05) | | | | 1.82 |

Note: Means with different superscripts are significantly different from one another ($P < 0.05$)

**Table 11.21:** Mean Chlorophyll content (SPAD).

| Microwave Treatment (J cm$^{-3}$) | Nitrogen Treatment (mg L$^{-1}$ of Solution) | | | |
|---|---|---|---|---|
| | 0 | 50 | 100 | 150 |
| 0 | 43.48$^{a}$ | 42.97$^{a}$ | 45.6$^{ab}$ | 46.27$^{ab}$ |
| 60 | 49.04$^{b}$ | 48.37$^{b}$ | 49.5$^{b}$ | 49.94$^{b}$ |
| LSD (P = 0.05) | | | | 2.79 |

Note: Means with different superscripts are significantly different from one another ($P < 0.05$)

**Table 11.22:** Mean plant biomass yield from three sampled plants per pot (g pot$^{-1}$).

| Microwave Treatment (J cm$^{-3}$) | Nitrogen Treatment (mg L$^{-1}$ of Solution) | | | |
|---|---|---|---|---|
| | 0 | 50 | 100 | 150 |
| 0 | 2.17$^{a}$ | 2.15$^{a}$ | 2.35$^{a}$ | 2.43$^{a}$ |
| 60 | 2.77$^{ab}$ | 3.12$^{bc}$ | 3.50$^{c}$ | 3.54$^{c}$ |
| LSD (P = 0.05) | | | | 0.67 |

Note: Means with different superscripts are significantly different from one another ($P < 0.05$)

**Table 11.23:** Mean Photosynthetic rate (µmole $CO_2$ $m^{-2}$ $s^{-1}$).

| Microwave Treatment ($J\ cm^{-3}$) | Nitrogen Treatment ($mg\ L^{-1}$ of Solution) | | | |
|---|---|---|---|---|
| | **0** | **50** | **100** | **150** |
| 0 | $10.77^{a}$ | $11.62^{a}$ | $11.85^{a}$ | $10.19^{a}$ |
| 60 | $17.52^{b}$ | $18.21^{b}$ | $18.73^{b}$ | $17.83^{b}$ |
| LSD (P = 0.05) | | | | 1.40 |

Note: Means with different superscripts are significantly different from one another (P < 0.05)

**Table 11.24:** Mean dry biomass at harvest (g $pot^{-1}$).

| Microwave Treatment ($J\ cm^{-3}$) | Nitrogen Treatment ($mg\ L^{-1}$ of Solution) | | | |
|---|---|---|---|---|
| | **0** | **50** | **100** | **150** |
| 0 | $19.21^{a}$ | $20.62^{ab}$ | $22.38^{b}$ | $21.54^{ab}$ |
| 60 | $28.89^{c}$ | $28.46^{c}$ | $29.72^{c}$ | $32.64^{d}$ |
| LSD (P = 0.05) - Microwave | | | | 2.94 |

Note: Means with different superscripts are significantly different from one another (P < 0.05)

**Table 11.25:** Mean grain yield at harvest (g $pot^{-1}$).

| Microwave Treatment ($J\ cm^{-3}$) | Nitrogen Treatment ($mg\ L^{-1}$ of Solution) | | | |
|---|---|---|---|---|
| | **0** | **50** | **100** | **150** |
| 0 | $8.43^{a}$ | $8.93^{a}$ | $9.74^{a}$ | $9.60^{a}$ |
| 60 | $12.95^{b}$ | $13.00^{b}$ | $12.97^{b}$ | $15.22^{c}$ |
| LSD (P = 0.05) - Microwave | | | | 1.60 |

Note: Means with different superscripts are significantly different from one another (P < 0.05)

# 12 A System Model for Crop Yield Potential as a Function of Microwave Weed Control over Time

## 12.1 Introduction

As pointed out in Chapter 3, equation (12.1) approximates the crop yield potential in response to weed infestation and herbicide application. This model also attempts to account for herbicide resistance within the weed population and the potential toxicity of the herbicide to the crop itself.

$$Y = Y_o \left\{ 1 - \frac{I \cdot [W(1-N-D_o)-E_m+I_m] \cdot \left[1 - S \cdot e^{\frac{-ag^2}{2}} + S \cdot e^{\frac{-ag^2}{2} - \lambda H}\right]}{100 \left\{ e^{ct}\left[1+e^{-\left(\frac{t-t_o}{d}\right)}\right] + \frac{I \cdot [W(1-N-D_o)-E_m+I_m] \cdot \left[1 - S \cdot e^{\frac{-ag^2}{2}} + S \cdot e^{\frac{-ag^2}{2} - \lambda H}\right]}{A_w} \right\}} + aH^2 - bH \right\} \quad (12.1)$$

Where I is the percentage yield loss as the weed density tends towards zero (= 0.38 (Bosnić and Swanton 1997)), W is the viable seed bank, N is the natural death rate for the whole population (Note: this is expressed as a fraction of the initial seed bank population Wo), $D_0$ is a fraction of the seed population developing dormancy (Note: this is expressed as a fraction of the initial seed bank population $W_0$, $E_m$ is the seed emigration out of the area of interest, $I_m$ is the seed immigration into the area of interest, S is the initial portion of the weed population that is susceptible to herbicides, s is the selection pressure for herbicide resistance in the system, g is the number of weed generations in the study period, c is the rate at which I approaches zero as time approaches ∞ (= 0.017 (Bosnić and Swanton 1997)), t is the time difference between crop emergence and weed emergence, $t_0$ is the time for 50 % germination of the viable seed bank, d is the slope of the seed bank recruitment curve at to, λ is the efficacy of the herbicide killing action, H is the herbicide dose, and Aw is the percentage yield loss as weed density approaches ∞ (= 38.0 (Bosnić and Swanton 1997)).

Using the same basic derivation, that was used to develop the herbicide transfer function response in equation (12.1), but substituting parameterised versions of the microwave weed responses derived from experimental data presented earlier instead of the herbicide efficacy components of equation (8), provides the relationship between crop yield potential and applied microwave energy:

$$Y = Y_o \left\{ 1 - \frac{I \cdot [W(1-N-D_o)-E_m+I_m] \cdot \{a \cdot erfc[b(\Psi - g)] + e \cdot erfc[f(\Psi - k)]\}}{100 \left\{ e^{ct}\left[1+e^{-\left(\frac{t-t_o}{d}\right)}\right] + \frac{I \cdot [W(1-N-D_o)-E_m+I_m] \cdot \{a \cdot erfc[b(\Psi - g)] + e \cdot erfc[f(\Psi - k)]\}}{Aw} \right\}} + l + m \cdot erf[n(\Psi - q)] \right\} \quad (12.2)$$

Where a, b, g, e, f, and k are constants derived from experimental data for different weed species described in Chapters 8 and 9. The parameters l, m, n and q are associated with the yield response in crops that was described in Chapter 11.

Differentiating equation (12.2) with respect to Ψ determines the sensitivity of crop yield to microwave weed and soil treatments:

$$\frac{dY}{d\Psi} = Y_o \frac{I\cdot[W(1-N-D_o)-E_m+I_m]\cdot\left\{\frac{2ab}{\sqrt{\pi}}\cdot e^{-[b^2(\Psi-g)^2]}+\frac{2ef}{\sqrt{\pi}}\cdot e^{-[f^2(\Psi-k)^2]}\right\}}{100\left\{e^{ct}\left[1+e^{-\left(\frac{t-t_o}{d}\right)}\right]+\frac{I\cdot[W(1-N-D_o)-E_m+I_m]\cdot\{a\cdot erfc[b(\Psi-g)]+e\cdot erfc[f(\Psi-k)]\}}{A_w}\right\}} - Y_o \frac{I^2\cdot[W(1-N-D_o)-E_m+I_m]^2\cdot\{a\cdot erfc[b(\Psi-g)]+e\cdot erfc[f(\Psi-k)]\}\cdot\left\{\frac{2ab}{\sqrt{\pi}}\cdot e^{-[b^2(\Psi-g)^2]}+\frac{2ef}{\sqrt{\pi}}\cdot e^{-[f^2(\Psi-k)^2]}\right\}}{100^2\left\{e^{ct}\left[1+e^{-\left(\frac{t-t_o}{d}\right)}\right]+\frac{I\cdot[W(1-N-D_o)-E_m+I_m]\cdot\{a\cdot erfc[b(\Psi-g)]+e\cdot erfc[f(\Psi-k)]\}}{A_w}\right\}^2} + Y_o \frac{2mn}{\sqrt{\pi}}\cdot e^{-[n^2(\Psi-q)^2]} \tag{12.3}$$

Equations (12.1) to (12.3) were coded into a simple cropping system model using the MatLab (version 2017a) software platform. Using data published by Bosnić and Swanton (1997) and Yin, et al. (2008) for some of the crop and weed parameters and assuming a seed mortality rate of 10% each year, the system transfer function was used to analyse crop yield potential as a function of applied microwave energy.

One possible scenario for using microwave energy in a broad acre cropping system is as a once off microwave soil treatment to deplete the weed seed bank, followed by a resumption of herbicide weed control. It has been shown that microwave soil treatment can destroy seeds in the top 5 cm of soil (Brodie, et al., 2007b; Brodie and Hollins, 2015). It is also apparent that 90% of the viable weed seed bank in zero-till systems can be found in the top 5 cm of soil (Swanton, et al., 2000); therefore, the impact of a once off microwave soil treatment can be estimated by comparing the time based crop response from a conventional herbicide regime with another analysis with an initial seed bank population of 10% of the original analysis.

Figure 12.1 shows the potential crop yield response to microwave-based weed control, as a function of applied microwave energy. This model implies that an improvement in normalised crop yield potential, above unity, may be possible, due to the enhanced crop yield in microwave treated soil. It is also important to understand that microwave soil treatment has the potential to deactivate the dormant weed seed bank in the upper layers of soil. It is unclear how the depletion of the soil seed bank may affect the longer-term potential of microwave weed control. Residual chemicals can provide some seedbank depletion; however chemical soil treatment often requires a delay before the treated site can be accessed or used. Unlike residual chemical options, microwave soil treatment is a purely thermal effect (Nelson, 1996), therefore the treated site is accessible as soon as the soil cools.

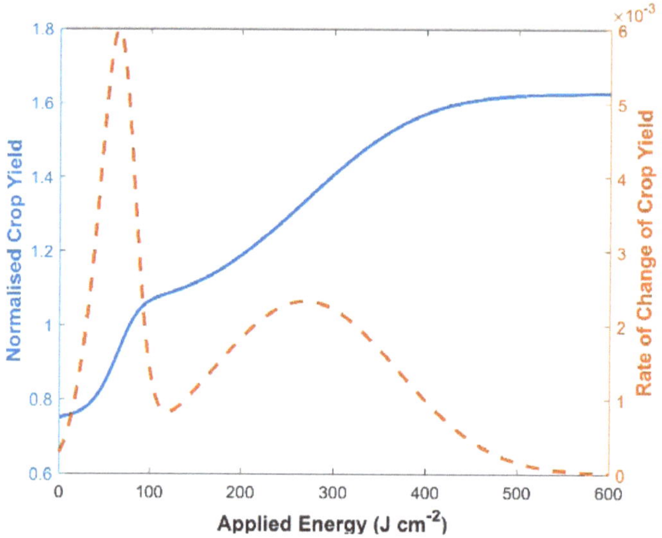


**Figure 12.1:** Relative rice crop yield as a function of applied microwave energy, based on the derived microwave response model in equations (12.2) and (12.3).

The mathematical system transfer function presented in this paper is useful for assessing the potential of using microwave weed management strategies as a tool for managing herbicide resistant weed populations.

Figure 12.2 shows the 15 year crop response (i.e. 15 annual generations of weeds) to ongoing herbicide weed management, assuming an initial weed seed bank density of 500 seeds $m^{-2}$, an initially small resistant population ($p_0 = 1 \times 10^{-8}$), an average seed set of 700 seeds per weed plant, a slightly positive selection coefficient of 0.0001 for herbicide resistance (Baucom and Mauricio, 2004), and other key herbicide data published by Bosnić and Swanton (1997) and Yin, et al. (2008). Figure 12.3 shows the 15-year crop response to the same ongoing herbicide weed management, except that the initial weed seed bank density is reduced to 50 seeds $m^{-2}$ to account for a once off microwave soil treatment. The difference in crop yield potential and soil seed bank growth is shown in Figure 12.4.

The cumulative yield advantage over the 15-year simulation, offered by a once off microwave soil treatment to deplete the weed seed bank, is equivalent to 1.5 full crops. When this is coupled with the 55% increase in crop yield potential in a single season due to microwave soil treatment, as demonstrated in the earlier chapters, the full advantage of a once off microwave soil treatment in a cropping system may be equivalent to 2.05 additional crops. Another advantage of depleting the initial soil weed seed bank is that the seed bank grows at a significantly slower rate than would otherwise occur.

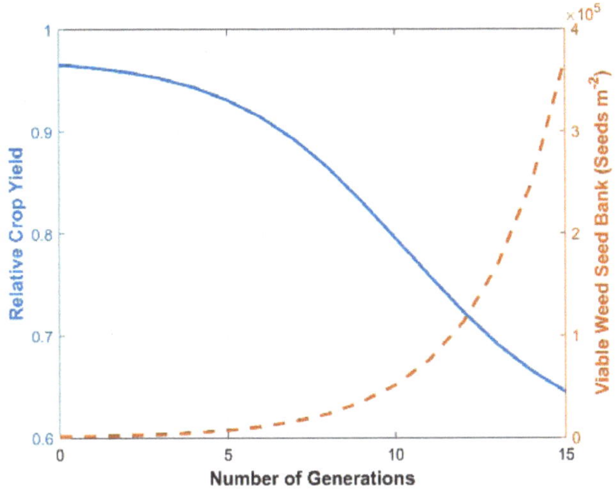


**Figure 12.2:** Modelling the generational impact of herbicide resistant weeds on potential crop yield under continuous herbicide weed management, based on equations (8) and (9).

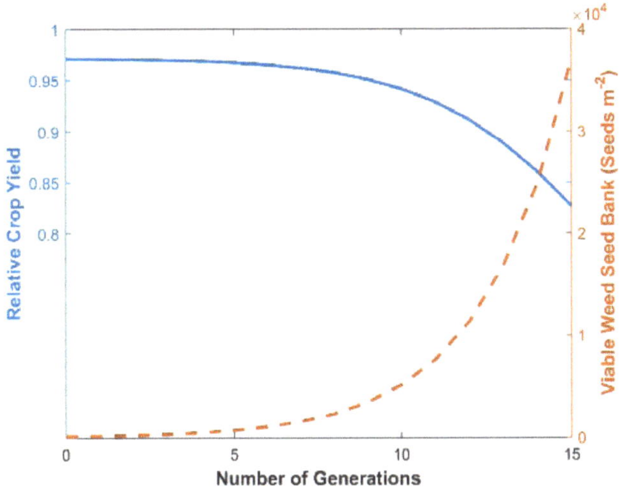


**Figure 12.3:** Modelling the generational impact of herbicide resistant weeds on potential crop yield assuming a 90% depletion of the weed seed bank by a once off microwave soil treatment.

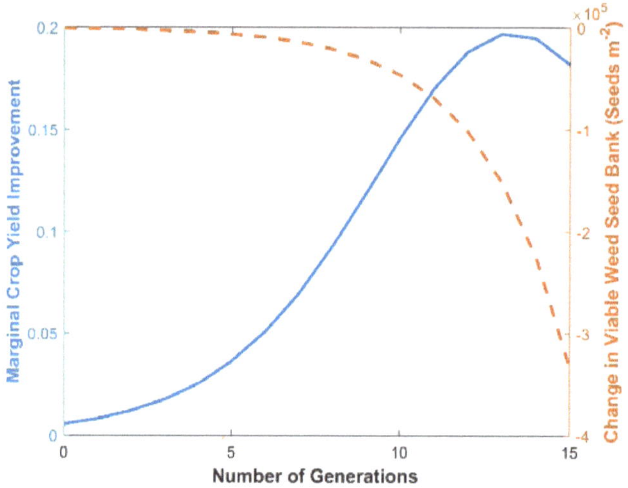


**Figure 12.4:** Difference in crop yield potential and cumulative soil seed bank between the scenarios depicted in Figure 12.2 and Figure 12.3.

It is also apparent from Figure 12.4 that the crop yield advantage has a limited life time and that after 15 years the difference in crop yield potential begins to decline. This suggests that a periodic application of microwave soil treatment to "restart" the conventional herbicide strategy may be a viable option.

For low yielding, low value crops, the expenditure needed to treat the soil with microwave energy may not be justified; however, for higher yielding, high value horticultural or rice crops, this expenditure may be more than balanced by the additional value derived by the yield advantage provided by a once off microwave treatment.

All modelling exercises are only indicative. The true value of microwave weed and soil treatment, if there is one, will only become evident as field experience with the technology over many years is gained; however, these models provide motivation to develop the technology to the point where field experience can be gained.

## 12.2 Conclusion

This chapter has developed a cropping system transfer function relating microwave application energy to potential crop yield. The resulting transfer function reveals the microwave weed and soil treatment has the potential to increase normalised crop yield potential above unity, resulting in significant increases in production potential. It also suggests that a once off microwave soil treatment to deplete the weed seed bank may offer long term yield advantages under conventional herbicide weed management scenarios.

## 12.3 References

Baucom, R. S. and Mauricio, R. (2004). Fitness costs and benefits of novel herbicide tolerance in a noxious weed. Proceedings of the National Academy of Sciences of the United States of America. 101(36), 13386-13390.

Bosnić, A. Č. and Swanton, C. J. (1997). Economic Decision Rules for Postemergence Herbicide Control of Barnyardgrass (Echinochloa crus-galli) in Corn (Zea mays). Weed Science. 45(4), 557-563.

Brodie, G., Hamilton, S. and Woodworth, J. (2007a). An assessment of microwave soil pasteurization for killing seeds and weeds. Plant Protection Quarterly. 22(4), 143-149.

Brodie, G. and Hollins, E. (2015). The Effect of Microwave Treatment on Ryegrass and Wild Radish Plants and Seeds. Global Journal of Agricultural Innovation, Research & Development. 2(1), 16-24.

Brodie, G., Pasma, L., Bennett, H., Harris, G. and Woodworth, J. (2007b). Evaluation of microwave soil pasteurization for controlling germination of perennial ryegrass (Lolium perenne) seeds. Plant Protection Quarterly. 22(4), 150-154.

Brodie, G., Ryan, C. and Lancaster, C. (2012a). The effect of microwave radiation on Paddy Melon (Cucumis myriocarpus). International Journal of Agronomy. 2012, 1-10.

Brodie, G., Ryan, C. and Lancaster, C. (2012b). Microwave technologies as part of an integrated weed management strategy: A Review. International Journal of Agronomy. 2012, 1-14.

Hollins, E. (2013). Microwave Weed Control and the Soil Biota. Unpublished thesis. Melbourne: University of Melbourne, Melbourne School of Land and Environment.

Nelson, S. O. (1996). A review and assessment of microwave energy for soil treatment to control pests. Transactions of the ASAE. 39(1), 281-289.

Swanton, C. J., Shrestha, A., Knezevic, S. Z., Roy, R. C. and Ball-Coelho, B. R. (2000). Influence of tillage type on vertical weed seedbank distribution in a sandy soil. Canadian Journal of Plant Science. 80(2), 455-457.

Yin, X. L., Jiang, L., Song, N. H. and Yang, H. (2008). Toxic Reactivity of Wheat (Triticum aestivum) Plants to Herbicide Isoproturon. Journal of Agricultural and Food Chemistry. 56(12), 4825-4831.

# 13 A Preliminary Economic Assessment of the Microwave Technology in an Herbicide Resistant World

## 13.1 Introduction

A system transfer function for crop yield potential as a function of herbicide application has been derived in Chapter 3. This transfer function can be used to determine the crop response to expenditure on weed management using herbicides. According to the Australian Bureau of Statistics (2010), the area under grain production was approximately 24 million ha in 2008. According to Jones, Vere, Alemseged, and Medd (2005), the total expenditure for weed management in cropping systems, including herbicide application or tillage, was AU$1.18 billion. Therefore, allowing for inflation (Tab. 13.1), the expenditure per ha for weed management in Australia was about AU$57 ha$^{-1}$, in 2010, when the area under cropping was determined.

Allowing for further inflation, the current average direct expenditure for weed control is therefore approximately AU$74.67 ha$^{-1}$. It was also pointed out in Chapter 2 that the indirect costs of herbicide use in Australia, due to environmental contamination, crop yield loss and human health costs, could be approximately US$433 million per annum. Allowing for inflation and the currency exchange rate of AU$1.00 = US$0.75, this equates to an additional AU$31.50 ha$^{-1}$; therefore, the real expenditure (both direct and indirect) of herbicide treatment could be up to AU$106.20 ha$^{-1}$.

## 13.2 Microwave Weed Control

In Chapter 12, a system transfer function for microwave weed management has also been derived from various experimental data reported here and elsewhere (Brodie, Bootes, & Reid, 2015; Brodie & Hollins, 2015). The transfer function for crop yield potential as a function of microwave weed treatment is:

$$Y = Y_o \left\{ 1 - \frac{I \cdot [W(1-N-D_O) - E_m + I_m] \cdot \{a \cdot erfc[b(\Psi - g)] + e \cdot erfc[f(\Psi - k)]\}}{100 \left\{ e^{ct} \left[ 1 + e^{-\left(\frac{t-t_O}{d}\right)} \right] + \frac{I \cdot [W(1-N-D_O) - E_m + I_m] \cdot \{a \cdot erfc[b(\Psi - g)] + e \cdot erfc[f(\Psi - k)]\}}{A_W} \right\}} + l + m \cdot erf[n(\Psi - q)] \right\} \tag{13.1}$$

This function uses many of the same parameters as the earlier transfer function derived for herbicide treatment (Brodie, 2014); however, it allows for the responses to applied microwave energy ($\Psi$) of emerged broad-leafed plants ($LD_1$), some hardier grasses ($LD_2$), and the enhanced crop yield and seed bank destruction associated with high microwave treatment energies ($LD_3$) (Brodie et al., 2015). This equation can also

**Table 13.1:** Annual inflation figures for Australia (Source: Australian Bureau of Statistics, 2010).

| Year | Annual Inflation Rate (%) | Cumulative Effect (Multiplier of 2005 costs) |
|---|---|---|
| 2017 | | |
| 2016 | 1.3 | 1.31 |
| 2015 | 1.5 | 1.30 |
| 2014 | 2.5 | 1.28 |
| 2013 | 2.5 | 1.25 |
| 2012 | 1.7 | 1.22 |
| 2011 | 3.3 | 1.19 |
| 2010 | 2.9 | 1.16 |
| 2009 | 1.7 | 1.12 |
| 2008 | 4.4 | 1.11 |
| 2007 | 2.3 | 1.06 |
| 2006 | 3.5 | 1.04 |
| 2005 | 2.7 | 1.00 |

be used to determine the Loss-Expenditure Frontier for microwave control.

In Chapter 8, the results of a travelling microwave trailer experiment were reported. In this experiment, 100% control of kikuyu grass was achieved with a travel speed of about 720 m $h^{-1}$. The applicator treats a strip about 150 mm wide and there are four microwave generators on the trailer; therefore, it can treat an area of 432 $m^2$ $h^{-1}$. The 7 kW electrical generators on the trailer have a specific fuel consumption of 2.0 L $h^{-1}$; therefore, with two electrical generators on the trailer, the fuel consumption is about 4.0 L, or $9.3 \times 10^{-3}$ L $m^{-2}$. Assuming a fuel price of AU$0.70 $L^{-1}$ for land holders, the cost of treatment is about AU$0.0065 $m^{-2}$. The trailer prototype is set up to demonstrate inter-row weed treatment in a crop. In this configuration, the costs of treatment are about AU$64.80 $ha^{-1}$.

A larger system, run from the PTO of a tractor could potentially perform better than the trailer prototype. Figure 13.1 compares microwave weed management to herbicide weed management, assuming a larger prototype system and engine performance based on data from Durković and Damjanović (2006). Because optimal travel speed and performance on the trailer system needs to be clarified, the data in Figure 13.1 should be regarded as indicative only; however, for inter-row weed control in crops, it appears that microwave weed management may be comparable in expenditure to herbicide weed management.

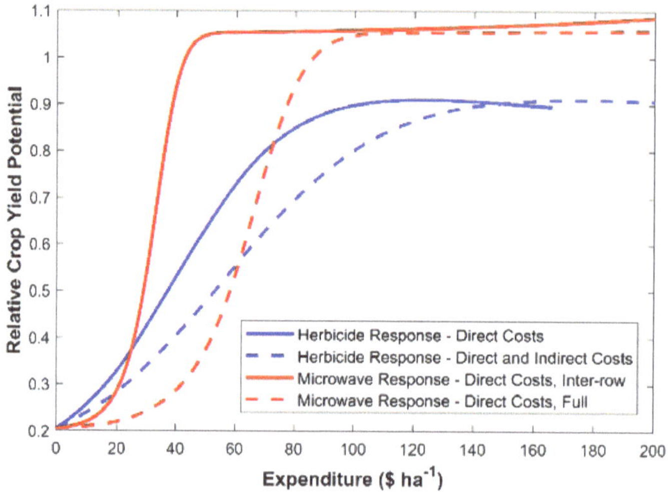


**Figure 13.1:** Indicative comparison of crop yield potential for microwave and herbicide based weed management systems, using the microwave energy to knock down weed plants.

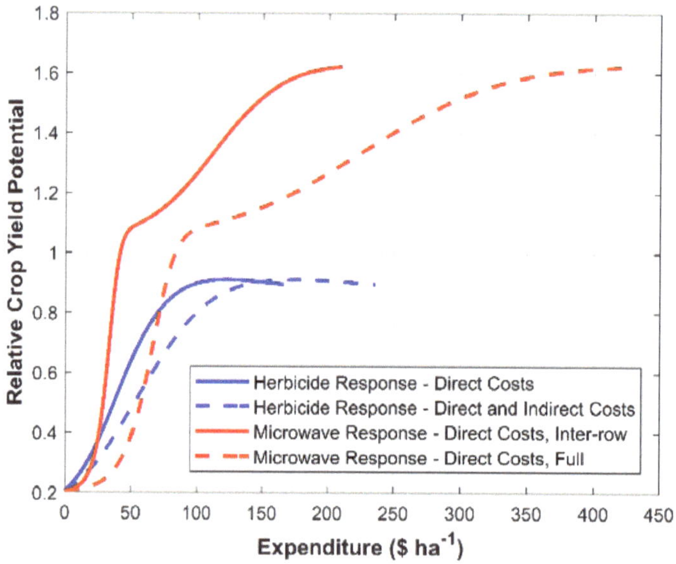


**Figure 13.2:** Indicative crop yield response to microwave soil treatment based on microwave soil treatment.

As pointed out earlier, microwave soil treatment has many secondary benefits and can be regarded as a soil fumigation treatment. Figure 13.2 shows indicative crop responses to expenditure on microwave soil treatment.

## 13.3 References

Australian Bureau of Statistics. (2010). Year Book Australia. Canberra: Australian Bureau of Statistics.
Brodie, G. (2014). Derivation of a Cropping System Transfer Function for Weed Management: Part 1 – Herbicide Weed Management. Global Journal of Agricultural Innovation, Research & Development, 1(1), 11-16.
Brodie, G., Bootes, N., & Reid, G. (2015). Invited Paper. Plant growth and yield of wheat and canola in microwave treated soil. Paper presented at the IMPI's 49th Microwave Power Symposium, San Diego, California, USA.
Brodie, G., & Hollins, E. (2015). The Effect of Microwave Treatment on Ryegrass and Wild Radish Plants and Seeds. Global Journal of Agricultural Innovation, Research & Development, 2(1), 16-24.
Durković, R., & Damjanović, M. (2006). Regression models of specific fuel consumption curves and characteristics of economic operation of internal combustion engines. Facta Universitatis, Mechanical Engineering, 4(1), 17-26.
Jones, R. E., Vere, D. T., Alemseged, Y., & Medd, R. W. (2005). Estimating the economic cost of weeds in Australian annual winter crops. Agricultural Economics, 32(3), 253-265. doi:10.1111/j.1574-0862.2005.00217.x

# 14 Industry Acceptance and Conclusion

## 14.1 Introduction

Historically, the adoption of new technology in established industries with mature infrastructures has been problematic. Existing systems, although they may not be as efficient or as effective as the newer technology, may be favoured because of: the long-term investment in the existing infrastructure; the challenge of integrating the new technology into the existing business; and a lack of understanding of the new technology and its potential by decision makers. Therefore, adoption of a new technology must be driven by the user and is often based on a clearly perceived need that can be better satisfied by the new technology rather than the existing infrastructure.

## 14.2 Safety

According to the Radiation Protection Standard for Maximum Exposure Levels (Australian Radiation Protection and Nuclear Safety Agency, 2002), the maximum time average operator exposure to electric fields at 2.45 GHz is 137 V $m^{-1}$ (RMS), while the exposure for the general public must be below 61.4 V $m^{-1}$ (RMS).

Given the modern reliance on Global Positioning Systems (GPS) and precision farming, it is also important that microwave weed systems do not interfere with these systems. The global positioning system (GPS) makes use of medium altitude satellites to determine position, velocity and time at the receiver. GPS receivers can access the L1 (1.575 GHz), L2 (1.227 GHz), and L5 (1.176 GHz) bands (Falade, et al., 2012). Antennae on GPS receivers vary in their configuration, but micro-strip "patch" antennae are becoming more common because of their low profile, light weight, low cost, ruggedness, and conformability (Chang, et al., 1986).

Patch antennae provide variable bandwidths. For example, the stacked patch antenna, designed by Falade, et al. (2012), provides 10 dB of attenuation outside its operating bandwidths for GPS L1, L2, and L5 frequency bands, which are 1.160–1.182, 1.214–1.232, and 1.568–1.598 GHz, respectively. At frequencies outside these ranges, the coupling of microwave fields into the GPS receiver is very low. Exposure of GPS systems to microwave fields, at frequencies other than those used by the GPS system, should be limited to the same levels as exposure to the general public.

When the trailer prototype system was tested using a Tenmars TM-194 microwave leakage detector, the time average field strength at the location of the operator for the trailer system was 13.2 V $m^{-1}$. The maximum measured field strength was 47.6 $Vm^{-1}$. This is well below the allowable exposure for both the operator and for the general public.

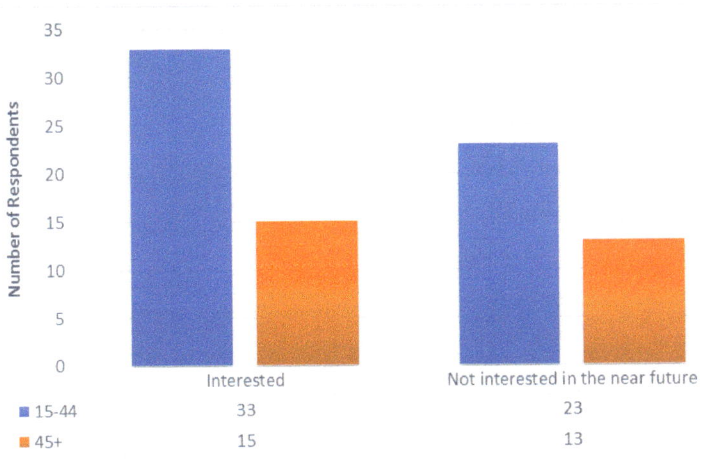


**Figure 14.1:** Response of rural community to their level of interest in trying microwave weed control technology, as a function of age.

The occasional burst of higher field strength is due to some channelling effects under the mental trailer and only extends a few centimetres above the ground.

The average field strength at the hitching point at the front of the trailer was 10.6 V $m^{-1}$. The maximum measured field strength was 32.5 V $m^{-1}$. This also is well below the allowable exposure for both the operator and for the general public, and indicates that the exposure levels of a GPS system, which is further away from the microwave system than the hitching point, will have minimal exposure risk.

## 14.3 Industry Acceptance

A questionnaire was developed to capture data from farming communities around Australia. Although there was a small bias in the responses due to age of the respondents, there was some support for trying the microwave technology among the younger members of the rural community (Figure 14.1). Some comments suggested that microwave weed and soil treatment could find its first commercial applications in high value horticulture, as a substitute for soil fumigation, along road sides, or in urban applications, where concerns about chemical exposure is growing.

## 14.4 Key Features of the Technology

In summary, microwave technology can be used to kill already emerged weed plants, like a knock-down herbicide application, or it can be applied to the soil, like a soil fumigant. Experimental data, which has been gathered during this research programme, indicates that weed plant treatment requires a smaller amount of microwave energy than soil treatment; however, soil treatment appears to provide several additional benefits, which killing emerged plants does not. These include: direct control of the weed seed bank; control of several potentially pathogenic organisms (nematodes and fungi); enhanced crop growth; and some residual effect over two or more seasons.

## 14.5 Summary

Interest in the effects of high frequency electromagnetic waves on biological materials dates back to the late 19th century (Ark and Parry, 1940), while interest in the effect of high frequency waves on plant material began in the 1920's (Ark and Parry, 1940). Many of the earlier experiments on plant material focused on the effect of radio frequencies (RF) on seeds (Ark and Parry, 1940). In many cases, short exposure resulted in increased germination and vigour of the emerging seedlings (Tran, 1979; Nelson and Stetson, 1985); however long exposure usually resulted in seed death (Ark and Parry 1940; Bebawi, et al., 2007; Brodie, et al., 2009).

Davis, et al. (1971; Davis, 1973) were among the first to study the lethal effect of microwave heating on seeds. They treated seeds, with and without any soil, in a microwave oven and showed that seed damage was mostly influenced by a combination of seed moisture content and the energy absorbed per seed. Other findings from the study by Davis, et al. (1971) suggested that both the specific mass and specific volume of the seeds were strongly related to a seed's susceptibility to damage by microwave fields (Davis, 1973). The association between the seed's volume and its susceptibility to microwave treatment may be linked to the "*radar cross-section*" (Wolf, et al., 1993) presented by seeds to propagating microwaves. Large radar cross-sections allow the seeds to intercept, and therefore absorb, more microwave energy.

Barker and Craker (1991) investigated the use of microwave heating in soils of varying moisture content (10-280 g water/kg of soil) to kill 'Ogle' Oat (*Avena sativa*) seeds and an undefined number of naturalised weed seeds present in their soil samples. Their results demonstrated that a seed's susceptibility to microwave treatment is entirely temperature dependent. When the soil temperature rose to 75°C there was a sharp decline in both oat seed and naturalised weed seed germination. When the soil temperature rose above 80°C, seed germination in all species was totally inhibited.

Several patents dealing with microwave treatment of weeds and their seeds have been registered (Haller, 2002; Clark and Kissell, 2003; Grigorov, 2003); however, none of these systems appear to have been commercially developed. This may be due to concerns about the energy requirements to manage weed seeds in the soil using microwave energy. In a theoretical argument based on the dielectric and density properties of seeds and soils, Nelson (Nelson, 1996) demonstrated that using microwaves to selectively heat seeds in the soil "*cannot be expected*". He concluded that seed susceptibility to damage from microwave treatment is a purely thermal effect, resulting from soil heating and thermal conduction into the seeds. This has been confirmed experimentally by Brodie, et al. (2007a).

Experience confirms that microwave energy can kill a range of weed seeds in the soil (Davis, et al., 1971; Davis 1973; Barker and Craker, 1991; Brodie, et al., 2009).

Pre-sowing microwave irradiation of soil minimises weed establishment (Davis, et al., 1971; Davis, 1973; Sartorato, et al., 2006; Brodie, et al., 2012; Brodie and Hollins, 2015). It can also destroy the weed reproductive plant parts and their seeds that are covered up by soil at a depth of several centimetres (Diprose, et al. 1984; Brodie, et al., 2007b). Wayland et al. (1973) treated wheat and radish seeds *in situ* at 25 mm depth and moisture content of soil was 6.5%. They found that microwave treatment was toxic to seeds with a threshold of 10 J $cm^{-2}$ of energy density. Increasing power density was more effective at reducing the germination percentage of seeds than simply increasing energy density (exposure time at a fixed power level) for some species.

Davis et al. (1971) conducted an experiment to evaluate the effect of microwave treatment on the seedling survival percentage of twelve species. They described that the 48 hour germinated seedling showed no survival after a short exposure of microwave energy and concluded that susceptibility of young seedlings to microwave heating was highly correlated with moisture content and absorption of energy. Menges and Wayland (1974) compared post-emergence herbicides (methazoal, propachlor and perfluidone) with microwave at energy density of 45 – 720 J $cm^{-2}$ for weed suppression in an onion crop. They reported that microwave (360 J $cm^{-2}$) irradiation significantly inhibited weeds establishment. Additionally, minimum crop injury was noted in the case of microwave treatment (18%) compared to herbicides application (85%).

The current study has also demonstrated that microwave weed control can be applied to emerged weeds. The energy required to achieve plant mortality is usually less than that needed to treat soil; however, soil treatment offers several additional advantages over simple weed killing. These include: enhanced crop growth; better nitrogen use efficiency; control of some pathogenic organisms; and higher yields.

## 14.6 General Conclusion

Microwave energy can kill weed plants and their seeds in the soil. Chapter 8 explored the energy required to kill weed plants and Chapter 9 explored the energy needed to treat the soil using microwave energy. In Chapter 4, a summary table of energy expenditure for different types of weed management was developed. Microwave weed and soil treatment energy can now be slotted into the summary table. Weed plant treatment is comparable with herbicide treatment, while microwave soil treatment is comparable with soil fumigation. Both have a place in an integrated weed management strategy.

**Table 14.1:** Updated summary of mean energy needs for various weed control strategies, on a per treatment basis.

| Weed or Soil Treatment System | Main Application | Mean energy Requirements for complete coverage (GJ ha$^{-1}$) |
|---|---|---|
| Steam soil treatment | Soil Fumigation | 1,190 |
| Laser | Weed Control | 233.5 |
| Steam – Weed control | Weed Control | 17.7 |
| Tillage | Weed Control | 17.5 |
| Hot water | Weed Control | 14.3 |
| Hand Hoeing | Weed Control | 10.8 |
| Microwave treatment – Soil treatment | Soil Fumigation | 10.0 |
| Ox drawn tillage | Weed Control | 5.2 |
| Flaming | Weed Control | 3.4 |
| IR | Weed Control | 3 |
| UV | Weed Control | 2.5 |
| Herbicide (Total energy, including embodied energy for manufacture and transport) | Weed Control | 1.4 |
| Microwave treatment – weed plant treatment | Weed Control | 1.3 |

## 14.7 References


Ark, P. A. and Parry, W. (1940). Application of High-Frequency Electrostatic Fields in Agriculture. The Quarterly Review of Biology. 15(2), 172-191.

Australian Radiation Protection and Nuclear Safety Agency. (2002). RPS3 Radiation Protection Standard for Maximum Exposure Levels to Radiofrequency Fields—3 kHz to 300 GHz.

Barker, A. V. and Craker, L. E. (1991). Inhibition of weed seed germination by microwaves. Agronomy Journal. 83(2), 302-305.

Bebawi, F. F., Cooper, A. P., Brodie, G. I., Madigan, B. A., Vitelli, J. S., Worsley, K. J. and Davis, K. M. (2007). Effect of microwave radiation on seed mortality of rubber vine (Cryptostegia grandiflora R.Br.), parthenium (Parthenium hysterophorous L.) and bellyache bush (Jatropha gossypiifolia L.). Plant Protection Quarterly. 22(4), 136-142.

Brodie, G., Botta, C. and Woodworth, J. (2007a). Preliminary investigation into microwave soil pasteurization using wheat as a test species. Plant Protection Quarterly. 22(2), 72-75.

Brodie, G., Hamilton, S. and Woodworth, J. (2007b). An assessment of microwave soil pasteurization for killing seeds and weeds. Plant Protection Quarterly. 22(4), 143-149.

Brodie, G., Harris, G., Pasma, L., Travers, A., Leyson, D., Lancaster, C. and Woodworth, J. (2009). Microwave soil heating for controlling ryegrass seed germination. Transactions of the American Society of Agricultural and Biological Engineers. 52(1), 295-302.

Brodie, G. and Hollins, E. (2015). The Effect of Microwave Treatment on Ryegrass and Wild Radish Plants and Seeds. Global Journal of Agricultural Innovation, Research & Development. 2(1), 16-24.

Brodie, G., Ryan, C. and Lancaster, C. (2012). The effect of microwave radiation on Paddy Melon (Cucumis myriocarpus). International Journal of Agronomy. 2012, 1-10.

Clark, W. J. and Kissell, C. W. (2003). System and Method for In Situ Soil Sterilization, Insect Extermination and Weed Killing. Patent No. 20030215354A1.

Davis, F. S., Wayland, J. R. and Merkle, M. G. (1971). Ultrahigh-Frequency Electromagnetic Fields for Weed Control: Phytotoxicity and Selectivity. Science. 173(3996), 535-537.

Davis, F. S., Wayland, J. R. and Merkle, M. G. (1973). Phytotoxicity of a UHF Electromagnetic Field. Nature. 241(5387), 291-292.

Diprose, M. F., Benson, F. A. and Willis, A. J. (1984). The Effect of Externally Applied Electrostatic Fields, Microwave Radiation and Electric Currents on Plants and Other Organisms, with Special Reference to Weed Control. Botanical Review. 50(2), 171-223.

Grigorov, G. R. (2003). Method and System for Exterminating Pests, Weeds and Pathogens. Patent No. 20030037482A1.

Haller, H. E. (2002). Microwave Energy Applicator. Patent No. 20020090268A1.

Menges, R. M. and Wayland, J. R. (1974). UHF electromagnetic energy for weed control in vegetables. Weed Science. 22(6), 584-590.

Nelson, S. O. (1996). A review and assessment of microwave energy for soil treatment to control pests. Transactions of the ASAE. 39(1), 281-289.

Nelson, S. O. and Stetson, L. E. (1985). Germination responses of selected plant species to RF electrical seed treatment. Transactions of the ASAE. 28(6), 2051-2058.

Sartorato, I., Zanin, G., Baldoin, C. and De Zanche, C. (2006). Observations on the potential of microwaves for weed control. Weed Research. 46(1), 1-9.

Tran, V. N. (1979). Effects of Microwave Energy on the Strophiole, Seed Coat and Germination of Acacia Seeds. Australian Journal of Plant Physiology. 6(3), 277-287.

Wayland, J. R., Davis, F. S. and Merkle, M. G. (1973). Toxicity of an UHF Device to Plant Seeds in Soil. (3), 161.

Wolf, W. W., Vaughn, C. R., Harris, R. and Loper, G. M. (1993). Insect radar cross-section for aerial density measurement and target classification. Transactions of the American Society of Agricultural and Biological Engineers. 36(3), 949-954.

# Table of Figures

**Figure 3.1:** Normalised crop yield (blue line) and rate of change of crop yield (orange line) as a function of applied herbicide energy, based equation (3.12) — **16**

**Figure 3.2:** Normalised crop yield (blue line) and rate of change of crop yield (orange line) as a function of time (generations of weeds), based on (3.12) — **17**

**Figure 3.3:** Response surface for potential crop yield as a function of both herbicide application and generational change in the weed population — **17**

**Figure 3.4:** Response of crop yield potential to the number of days between weed emergence and crop emergence — **18**

**Figure 4.1:** Relationship between Travel Reduction Ratio and Traction Efficiency for a MF 6270 Massy-Ferguson tractor (tested at the Dookie Campus of Melbourne University) — **24**

**Figure 4.2:** Radiative spectral density at different temperatures as a function of temperature and wavelength — **30**

**Figure 5.1:** The electromagnetic spectrum, showing the microwave range in further detail — **42**

**Figure 5.2:** Dielectric constant and dielectric loss of water between 0°C and 100°C, the arrows showing the effect of increasing temperature — **44**

**Figure 5.3:** Dielectric constant and Loss Factor at 2.45 GHz for different parts per thousand w/w (ppt) salinity for the range for -20°C ~ +40°C — **45**

**Figure 5.4:** Dielectric properties of vegetative materials as a function of frequency and moisture content — **45**

**Figure 5.5:** Temperature distribution in (left) cross section of a rectangular block and (right) cross section of a sphere with dielectric properties of e' = 9.6 and e" = 2.5 after 120 seconds of microwave heating at 2.45 GHz, based on models derived by Brodie (2008) — **46**

**Figure 5.6:** Example of thermal runaway in plant material dur to changing dielectric properties as the sample dries out — **47**

**Figure 5.7:** Heating rate in 25 mm by 25 mm Eucalyptus regnans samples, heated in a wave-guide, as a function of applied power (Dotted line represents the response that constant diffusion should produce) — **49**

**Figure 6.1:** The Zapper III microwave prototype during field trials (Source: Anonymous, 1975) — **53**

**Figure 6.2:** Response of 11 species of weed to microwave energy (Sources: Menges & Wayland, 1974; Wayland et al., 1975) — **53**

**Figure 6.3:** Response of grasses (blue) and broad-leafed weeds (red) to microwave energy (Sources: Menges & Wayland, 1974; Wayland et al., 1975) — **54**

**Figure 6.4:** Response weed seeds in the soil to microwave energy, as a function of applied energy density, burial depth and irrigation status (Sources: Menges & Wayland, 1974; Wayland et al., 1975) — **55**

**Figure 7.1:** Dielectric properties of a clay based soil as a function of frequency and soil moisture (Data from: Wang, 1980) — **64**

**Figure 7.2:** Dielectric properties of a sand based soil as a function of frequency and soil moisture (Data from: Wang, 1980) — **64**

**Figure 7.3:** Dielectric properties of a sand (dotted) and clay (solid) as a function of soil moisture, at 2.45 GHz (Data from: Wang, 1980) — **65**
**Figure 7.4:** Geometry of a horn antenna showing a propagating wave front in the flare of the antenna (Nikolova, 2012) — **69**
**Figure 7.5:** Comparison of estimated relative instantaneous field strength along the centre line in front of the horn antenna as a function of distance from the antenna's aperture plane — **69**
**Figure 7.6:** Estimated microwave field strength at ground level from a 2 kW microwave system as a function of height of the 110 mm by 55 mm antenna above the ground — **70**
**Figure 7.7:** Trailer mounted four by **2** kW microwave weed killer — **71**
**Figure 7.8:** Calculated temperature change distribution in clay soil after 30 seconds of heating using the 2 kW microwave system and a horn antenna with an aperture of 110 mm by 55 mm at a range of 4 cm above the soil surface — **72**
**Figure 7.9:** Comparison of temperature profile as predicted by equation (7.8), indicated by the mesh, and the measured temperature profile, indicated by the point data, after heating sand for 240 s using a 1 kW microwave oven prototype system with a 180 mm by 90 mm horn antenna — **72**
**Figure 7.10:** Estimated change in soil temperature from the 2 kW microwave system after (a) 10 seconds with antenna at 18 cm above the soil, (b) 30 seconds with antenna at 18 cm above the soil, (c) 10 seconds with antenna at 2 cm above the soil, and (d) 30 seconds with antenna at 2 cm above the soil — **73**
**Figure 7.11:** Comparison of temperature change profile after 60 seconds (at 2.45 GHz) as predicted by equation (7.8) in dry soil (10% MC) left and moist soil (35% MC) on the right — **74**
**Figure 7.12:** Comparison of temperature change profile after 90 seconds of microwave heating, at 2.45 GHz, in sandy soil (left), compared with clay soil (right). — **74**
**Figure 7.13:** Comparison of temperature change in clay soil (left) treated using 922 MHz frequency and (right) treated using 2450 MHz frequency — **74**
**Figure 8.1:** Prototype system, fed from the magnetron of a 800 W microwave oven (Source: Bebawi et al., 2007) — **83**
**Figure 8.2:** Rendering of the second prototype system, with a 900 W magnetron. The magnetron is wired into the voltage doubler circuit of a normal microwave oven via a 2 m long cable. — **84**
**Figure 8.3:** Prototype four by 2 kW microwave weed killer (left) with details of the control panel (right) — **84**
**Figure 8.4:** Dose response curves for microwave treatment of four species of weed plant using a horn antenna. Dose curves are calculated using equations (8.1) and (8.2). — **86**
**Figure 8.5:** Thermal image of treated strip (Treatment = 1.5 m in 7 seconds), immediately after treatment — **87**
**Figure 8.6:** Visible light image of the treated strip, shown in Figure 8.5 — **88**
**Figure 8.7:** Image of four treated strips of kikuyu grass, taken (left) 4 days after treatment and (right) 20 days after treatment — **88**
**Figure 9.1:** Dose responses of ryegrass and wild oats seeds as a function of soil moisture, microwave energy at ground level and burial depth in soil — **92**
**Figure 10.1:** Normalised E. coli counts as a function of microwave energy and soil depth. — **103**

**Figure 10.2:** Combined response of bacteria to microwave treatment, as a function of applied energy density and soil depth (Sources of some data: Brodie et al., 2015; Cooper & Brodie, 2009) — **105**
**Figure 10.3:** Mean fungal counts across all treatment combinations in experiments by Vela et al. (1976). Error bars indicate Least Significant Differences (P = 0.05). — **106**
**Figure 10.4:** Mean fungal counts across all treatment combinations, as a function of soil depth, in experiments by Vela et al. (1976). Error bars indicate Least Significant Differences (P = 0.05). — **107**
**Figure 10.5:** Thermal images of the soil surface during microwave treatment using the prototype trailer — **110**
**Figure 10.6:** Fatal Temperature – Time curves for: Root knot nematodes, two species of bacteria, plant parts, and two species of fungi — **113**
**Figure 11.1:** Mean wheat plant height as a function of time since planting and microwave soil treatment energy (error bars represent LSD for P = 0.05) (Modified from: Brodie, Bootes, & Reid, 2015) — **117**
**Figure 11.2:** Comparison of wheat and canola plant growth as a function of microwave treatment energy (Control on the left and highest treatment on the right) — **117**
**Figure 11.3:** Comparison of rice plant growth as a function of microwave treatment energy (Control on the left and highest microwave treatment on the right) — **117**
**Figure 11.4:** Comparison of reproductive development in response to pre-sowing microwave soil heating in wheat crop at nitrogen treatment N2, in untreated control pot (left) and microwave treated pot (right). Adopted from Khan, Brodie, and Gupta (2016). — **121**
**Figure 11.5:** Biomass yield (g pot-1) of Wheat grown in a glasshouse as a function microwave (MW) treated soil (T0 = untreated (control); T1 = 120 sec 2.45GHz MW treatment under different rates of N application (N0 = 0 mg pot-1; N1= 22.5 mg pot-1; N2 = 45.0 mg pot-1; N3 = 67.5 mg pot-1) measured at harvest. Error bars indicate LSD at 5%. — **121**
**Figure 11.6:** Change in mean wheat grain yield induced by microwave soil treatment, compared to the equivalent control treatment, as a function of applied nitrogen — **123**
**Figure 11.7:** Comparison of microwave plants growing in soil that was treated with microwave energy at the beginning of the previous wheat crop (three pots on the left) and plants growing in soil from the previous crop that was not treated with microwave energy (three pots on the right) — **125**
**Figure 11.8:** Effect of microwave soil heating on an abundance of ammonia oxidizer archaea after 300 days of treatment — **125**
**Figure 11.9:** Effect of microwave soil heating on an abundance of ammonia oxidizer bacteria after 300 days of treatment — **126**
**Figure 11.10:** Effect of microwave soil heating on nitrogen derived from applied fertilizer in wheat crop under two years of microwave-nitrogen experiments — **128**
**Figure 11.11:** Grain yield as a function of above ground 15N at plant maturity — **129**
**Figure 11.12:** Experimental layout of the two microwave treatments: untreated control (T0) and MW treated (T1) of the Rice field experiment located at The University of Melbourne, Dookie College Campus in North East Victoria in 2015 – 16. — **130**
**Figure 11.13:** Prototype two horn antennae microwave system operated through two 2.45 GHz domestic microwave ovens (EMS8586V; Sanyo; Tokyo, Japan) used for soil irradiation in field conditions — **131**

**Figure 11.14:** Infrared thermal image of the microwave heating pattern of the soil in field conditions for the rice experiment, irradiated through the prototype microwave system (Figure 11.13). — **131**
**Figure 11.15:** Comparison of randomly sampled rice plants grown in the microwave treated plots (left) with rice plants grown in the control plots (right) — **132**
**Figure 11.16:** Comparison of microwave treated plot (left) with control plot (right) — **133**
**Figure 11.17:** Comparison of early growth establishment of rice crop. Plants on left collected from microwave treated plot and plants on right collected from untreated control plot. (Left image taken from Dookie Trial Site and right image taken from Old Coree, Jerilderie site.). — **134**
**Figure 11.18:** Experimental layout of the two microwave treatments: untreated control (T0) and MW treated (T1) of the Tomato crop located at Corop in North East Victoria in 2015 – 16 — **135**
**Figure 11.19:** Comparison of weed seedling establishment in microwave treated plot (left) and untreated control plot (right) in dryland wheat at Dookie — **137**
**Figure 11.20:** Plant height (cm) of Wheat grown in a glasshouse as function microwave (MW) treated soil ($T_0$ = untreated (control); $T_1$ = 120 sec 2.45GHz MW treatment under different rates of N application ($N_0$ = 0 mg pot$^{-1}$; $N_1$= 22.5 mg pot$^{-1}$; $N_2$ = 45.0 mg pot$^{-1}$; $N_3$= 67.5 mg pot$^{-1}$) measured at (A) 20 days after sowing (DAW); (B) 40 DAW; and (C) 60 DAW. Error bars indicate LSD at 5%. — **142**
**Figure 11.21:** Chlorophyll content (SPAD) of Wheat grown in a glasshouse as function microwave (MW) treated soil ($T_0$ = untreated (control); $T_1$ = 120 sec 2.45GHz MW treatment under different rates of N application ($N_0$ = 0 mg pot$^{-1}$; $N_1$= 22.5 mg pot$^{-1}$; $N_2$ = 45.0 mg pot$^{-1}$; $N_3$= 67.5 mg pot$^{-1}$) measured at (A) 20 days after sowing (DAW); (B) 40 DAW; and (C) 60 DAW. Error bars indicate LSD at 5%. — **143**
**Figure 11.22:** Grain yield per pot (g pot$^{-1}$) of Wheat grown in a glasshouse as a function microwave (MW) treated soil ($T_0$ = untreated (control); $T_1$ = 120 sec 2.45GHz MW treatment under different rates of N application ($N_0$ = 0 mg pot$^{-1}$; $N_1$= 22.5 mg pot$^{-1}$; $N_2$ = 45.0 mg pot$^{-1}$; $N_3$= 67.5 mg pot$^{-1}$) measured at harvest. Error bars indicate LSD at 5%. — **144**
**Figure 11.23:** Spikelets per spike of Wheat grown in a glasshouse as a function microwave (MW) treated soil ($T_0$ = untreated (control); $T_1$ = 120 sec 2.45GHz MW treatment under different rates of N application ($N_0$ = 0 mg pot$^{-1}$; $N_1$= 22.5 mg pot$^{-1}$; $N_2$ = 45.0 mg pot$^{-1}$; $N_3$= 67.5 mg pot$^{-1}$) measured at harvest. Error bars indicate LSD at 5%. — **144**
**Figure 11.24:** Grains per spikelet of Wheat grown in a glasshouse as a function microwave (MW) treated soil ($T_0$ = untreated (control); $T_1$ = 120 sec 2.45GHz MW treatment under different rates of N application ($N_0$ = 0 mg pot$^{-1}$; $N_1$= 22.5 mg pot$^{-1}$; $N_2$ = 45.0 mg pot$^{-1}$; $N_3$= 67.5 mg pot$^{-1}$) measured at harvest. Error bars indicate LSD at 5%. — **145**
**Figure 11.25:** Harvest Index (grain yield/biomass yield*100) of Wheat grown in a glasshouse as a function microwave (MW) treated soil ($T_0$ = untreated (control); $T_1$ = 120 sec 2.45GHz MW treatment under different rates of N application ($N_0$ = 0 mg pot$^{-1}$; $N_1$= 22.5 mg pot$^{-1}$; $N_2$ = 45.0 mg pot$^{-1}$; $N_3$= 67.5 mg pot$^{-1}$) measured at harvest. Error bars indicate LSD at 5%. — **145**
**Figure 11.26:** Tillers per shoot of Wheat grown in a glasshouse as a function microwave (MW) treated soil ($T_0$ = untreated (control); $T_1$ = 120 sec 2.45GHz MW treatment under different rates of N application ($N_0$ = 0 mg pot$^{-1}$; $N_1$= 22.5 mg pot$^{-1}$; $N_2$ = 45.0 mg pot$^{-1}$; $N_3$= 67.5 mg pot$^{-1}$) measured at harvest. Error bars indicate LSD at 5%. — **146**

**Figure 11.27:** Internodal distance (mm) of Wheat grown in a glasshouse as a function microwave (MW) treated soil ($T_0$ = untreated (control); $T_1$ = 120 sec 2.45GHz MW treatment under different rates of N application ($N_0$ = 0 mg pot$^{-1}$; $N_1$= 22.5 mg pot$^{-1}$; $N_2$ = 45.0 mg pot$^{-1}$; $N_3$ = 67.5 mg pot$^{-1}$) measured at harvest. Error bars indicate LSD at 5%. — **146**

**Figure 11.28:** Stem diameter (mm) of Wheat grown in a glasshouse as a function microwave (MW) treated soil ($T_0$ = untreated (control); $T_1$ = 120 sec 2.45GHz MW treatment under different rates of N application ($N_0$ = 0 mg pot$^{-1}$; $N_1$= 22.5 mg pot$^{-1}$; $N_2$ = 45.0 mg pot$^{-1}$; $N_3$ = 67.5 mg pot$^{-1}$) measured at harvest. Error bars indicate LSD at 5%. — **147**

**Figure 11.29:** Effect of Microwave and various level of Nitrogen on Plant Height of Wheat (40 & 60 DAS). — **147**

**Figure 12.1:** Relative rice crop yield as a function of applied microwave energy, based on the derived microwave response model in equations (12.2) and (12.3) — **154**

**Figure 12.2:** Modelling the generational impact of herbicide resistant weeds on potential crop yield under continuous herbicide weed management, based on equations (8) and (9) — **155**

**Figure 12.3:** Modelling the generational impact of herbicide resistant weeds on potential crop yield assuming a 90 % depletion of the weed seed bank by a once off microwave soil treatment — **155**

**Figure 12.4:** Difference in crop yield potential and cumulative soil seed bank between the scenarios depicted in Figure 12.2 and Figure 12.3 — **155**

**Figure 13.1:** Indicative comparison of crop yield potential for microwave and herbicide based weed management systems, using the microwave energy to knock down weed plants — **160**

**Figure 13.2:** Indicative crop yield response to microwave soil treatment based on microwave soil treatment — **160**

**Figure 14.1:** Response of rural community to their level of interest in trying microwave weed control technology, as a function of age —**163**

# Index


annual ryegrass · 85, 87, 92, 93, 96
applicator · 26, 28, 29, 31, 41, 46, 63, 64, 66, 67, 76, 86, 88, 89, 118, 161
archaea · 2, 103, 126
bacteria · 2, 103, 104, 105, 106, 108, 113, 114, 126, 128
barnyard grass · 1, 14, 85, 87
crop yield · 11, 12, 15, 16, 18, 57, 83, 118, 138, 140, 155, 156, 157, 158, 159, 161
destruction · 32, 161
dielectric properties · 30, 43, 44, 47, 63, 66, 74
electric field · 41, 42, 44, 45, 46, 63, 66, 67, 94, 165
fleabane · 85
*frequency* · 41, 42, 43, 52, 56, 63, 65, 66, 71, 74, 76, 83, 106, 132, 165, 166
fungi · 52, 103, 108, 111, 112, 113, 114, 115, 166
glyphosate · 5, 6, 7, 14, 18, 22, 27
herbicide resistance · 2, 6, 7, 11, 13, 15, 18, 22, 35, 57, 155, 157
herbicides · 1, 4, 5, 6, 7, 8, 11, 12, 13, 15, 22, 25, 138, 155, 161, 167
horn antenna · 63, 67, 68, 69, 71, 72, 75, 76, 84, 86, 92, 99, 104, 105, 118, 132, 136
insects · 1, 7, 8, 92, 103
magnetron · 41, 84, 85, 92
microwave effect · 48, 49
microwave fields · 41, 57, 63, 66, 71, 74, 83, 103, 165, 167
modelling · 11, 14, 99, 159
nematode · 52, 92, 103, 112, 114, 115, 118, 166
nitrogen · 2, 103, 121, 122, 123, 125, 126, 128, 130, 132, 139, 140, 168
novel microwave applicator · 57, 88, 90
pathogens · 1, 7, 58, 92
pesticides · 1
plant growth · 5, 32, 83, 114, 119, 121, 122, 124, 126, 129, 136, 138
prototype · 52, 56, 71, 84, 85, 88, 93, 104, 105, 107, 111, 132, 135, 162, 165
rice · 1, 2, 118, 120, 131, 132, 135, 136, 138, 141, 159
soil biota · 92, 103, 104, 105, 106, 107, 108, 114, 121
soil fumigation · 28, 36, 92, 95, 103, 118, 136, 140, 162, 166, 168
steam explosions · 88
steam soil treatment · 27, 28, 168
steam weed treatment · 22, 27, 28, 54, 94, 95
temperature distribution · 46, 71, 74, 94, 99
thermal images · 88, 111, 132
thermal weed control · 25, 26
tiller density · 119, 124, 132, 133, 134, 135, 138, 139, 149
transfer function · 11, 14, 15, 16, 155, 156, 159, 161
vigour · 52, 56, 139, 166
wave guide · 65, 67, 74, 84
weed seed bank · 2, 92, 156, 157, 158, 159, 166
wheat · 6, 56, 103, 118, 120, 121, 122, 128, 138, 139, 167
wild oats · 93, 103
Wild Radish · 87, 92, 93, 94, 96
Zapper · 52, 107

Kolibri 360 GmbH
Ferdinand-Jühlke-Straße 7
99095 Erfurt, Deutschland
produktsicherheit@kolibri360.de